Tools in Artificial Intelligence

Tools in Artificial Intelligence

Editor

Bilroy Muller

scitus academics

Tools in Artificial Intelligence
Edited by **Bilroy Muller**

ISBN: 978-1-68117-587-4
Library of Congress Control Number: 2016946442

www.scitusacademics.com

Notice

Printed in United States of America on Acid Free Paper

Preface

Artificial intelligence (AI) is the simulation of human intelligence processes by machines, especially computer systems. These processes include learning (the acquisition of information and rules for using the information), reasoning (using the rules to reach approximate or definite conclusions), and self-correction. Particular applications of AI include expert systems, speech recognition, and machine vision. There are a large number of tools used in AI, including versions of search and mathematical optimization, logic, methods based on probability and economics, and many others. The AI field is interdisciplinary, in which a number of sciences and professions converge, including computer science, mathematics, psychology, linguistics, philosophy and neuroscience, as well as other specialized fields such as artificial psychology. Today AI techniques have become an essential part of the technology industry, providing the heavy lifting for many of the most challenging problems in computer science.

Tools in Artificial Intelligence emphasize on some potential applications and give a partial picture of the current state-of-the-art of AI. The book provides technical aspects of specifying, developing, and evaluating the theoretical underpinnings and applied mechanisms of AI tools. Furthermore, it is convenient to inspire some future research ideas by identifying potential research directions. It is devoted to students, researchers and practitioners in this area or in related fields.

Table of Contents

Chapter 1

Metalearning: A survey of trends and technologies

Christiane Lemke[2], Marcin Budka[1,*], Bogdan Gabrys [1]

[1] Bournemouth University, Poole House, Talbot Campus, Fern Barrow, BH12 5BB Poole, UK

[2] Unister GmbH, Leipzig, Germany

ABSTRACT

Metalearning attracted considerable interest in the machine learning community in the last years. Yet, some disagreement remains on what does or what does not constitute a metalearning problem and in which contexts the term is used in. This survey aims at giving an all-encompassing overview of the research directions pursued under the umbrella of metalearning, reconciling different definitions given in scientific literature, listing the choices involved when designing a metalearning system and identifying some of the future research challenges in this domain.

KEYWORDS

Metalearning, Metaknowledge, extraction, Life-long learning

1. INTRODUCTION

The term *metalearning* first occurred in the area of educational psychology. One of the most cited researchers in this field, John Biggs, described metalearning as *being aware of and taking control of one's own learning* (Biggs 1985). Hence, metalearning is viewed as an understanding and adaptation of learning itself on a higher level than merely acquiring subject knowledge. In that way, a person aware and capable of metalearning is able to assess his or her learning approach and adjust it according to the requirements of a specific task.

Metalearning as used in a machine learning context has many similarities to this description. Subject knowledge translates into base-learning, where experience is accumulated for one specific learning task. Metalearning starts at a higher level and is concerned with accumulating experience over several applications of a learning system according to Brazdil et al. (2009).

In the last 20 years, machine learning research was faced with an increasing number of available algorithms including a multitude of parametrisation, preprocessing and postprocessing approaches as well as a substantially extended range of applications due to increasing computing power and wider availability of computer-readable data sets. By promoting a better understanding of machine learning itself, metalearning can provide an invaluable help avoiding extensive trial and error procedures for algorithm selection, and brute force searches for suitable parametrisation. Looking at how to profit from past experience of a predictive model on certain tasks can enhance the performance of a learning algorithm and allow to better understand what makes a given algorithm perform well on a given problem.

The idea of metalearning is not new, one of the first and seminal contributions having been provided by Rice (1976). However, the literal term only started appearing in machine learning literature in the 1990s, yet still many publications deal with problems related to metalearning without using the actual word. This contribution tries to grasp every point of view metalearning has been investigated from, citing books, research and review papers of the last decade. We hope this survey will provide a useful resource for the data mining and machine learning community.

The remainder of this paper is organized as follows. In Sect. 2 we review definitions of metalearning given in scientific literature, focusing on common themes occurring in all of them. Section 3 describes different notions of metalearning, linking them to the definitions given in Sect. 2. In Sect. 4 practical considerations arising when designing a metalearning system are discussed, while open research directions are listed in Sect. 5.

2. DEFINITION

In the 1990s, the term metalearning started to appear in machine learning research, although the concept itself dates back to the mid-1970s (Rice 1976). A number of definitions of metalearning have been given, the following list cites the main review papers and books from the last decade:

1) Metalearning studies how learning systems can increase in efficiency through experience; the goal is to understand how learning itself can become flexible according to the domain or task under study (Vilalta and Drissi 2002a).

2) The primary goal of metalearning is the understanding of the interaction between the mechanism of learning and the concrete contexts in which that mechanism is applicable (Giraud-Carrier 2008).

3) Metalearning is the study of principled methods that exploit metaknowledge to obtain efficient models and solutions by adapting machine learning and data mining processes (Brazdil et al. 2009).

4) Metalearning monitors the automatic learning process itself, in the context of the learning problems it encounters, and tries to adapt its behaviour to perform better (Vanschoren 2010).

Learning systems that adapt and improve by experience are a key concept of definitions 1, 3 and 4. This in itself however does not suffice as a description, as it basically applies to all machine learning algorithms. Metalearning becomes metalearning by looking at different problems, domains, tasks or contexts or simply past experience. This aspect is inherent in all of the definitions, although somewhat disguised in definition 3 using the term *metaknowledge* instead. Metaknowledge as described by the authors stands for knowledge to be exploited from past learning tasks, which may both mean past learning tasks on the same data or using data of another problem domain. Definition 2 differs in emphasising a better comprehension of the interaction between domains and learning mechanisms, which does not necessarily imply the goal of improved learning systems, but the pursuit of a better understanding of for which tasks individual learners succeed or fail.

Rephrasing, the common ground the above definitions share, we propose to define a metalearning system as follows:

Definition 1

1) A metalearning system must include a learning subsystem, which adapts with experience.

2) Experience is gained by exploiting metaknowledge extracted

   a) in a previous learning episode on a single dataset, and/or

   b) from different domains or problems.

Furthermore, a concept often used in metalearning is that of a bias, which, in this context, refers to a set of assumptions influencing the choice of hypotheses for explaining the data. Brazdil et al. (2009) distinguishes*declarative bias* specifying the representation of the space of hypotheses (for example representing hypotheses using neural networks only) and *procedural bias*, which affects the ordering of the hypotheses (for example preferring hypothesis with smaller runtime). The bias in base-learning according to this theory is fixed, whereas metalearning tries to choose the right bias dynamically.

3. NOTIONS OF METALEARNING

Metalearning can be employed in a variety of settings, with a certain disagreement in literature about what exactly constitutes a metalearning problem. Different notions will be presented in this section while keeping an eye on the question if they can be called metalearning approaches according to Definition 1. Figure 1groups general machine and metalearning approaches in relation to Definition 1. Each of the three circles presents a cornerstone of the definition (1: adapt with experience, 2a: meta-knowledge on same data set, 2b: meta-knowledge from different domains), the approaches are arranged into the circles and their overlapping sections depending on which parts of the definition applies to them. As an example, ensemble methods do generally work with experience gained with the same data set (definition 2a) and adapt with experience (definition 1), however, the only approach potentially applying all three parts of the definition is algorithm selection, which appears where all three circles overlap.

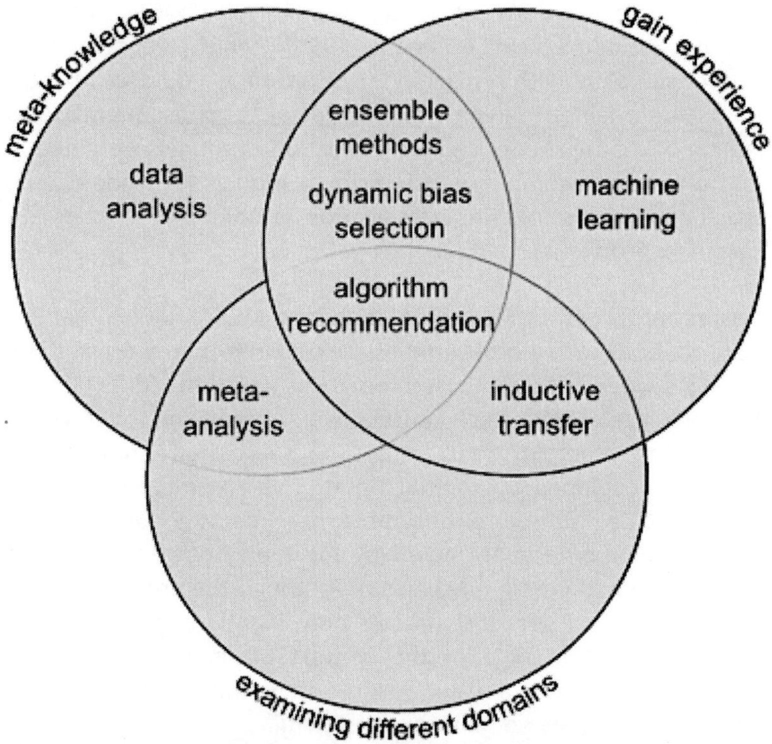

Figure 1. Notions of metalearning versus components of a metalearning system

3.1 Ensemble methods and combinations of base-learners

Model combination is often used when several applicable algorithms for a problem are available. Instead of selecting a single algorithm for a problem, the risk of choosing the wrong one can be reduced by combining all or a subset of the available outcomes. In machine learning, advanced model combination can be facilitated by ensemble learning according to Dietterich (2000) and Yao and Islam (2008), which comprises strategies for training and combining outputs of a number of machine learning algorithms. One often used approach of this type is resampling, leading to a number of ensemble generation techniques. Two very popular resampling-based ensemble building methods are:

- *Bagging* introduced in Breiman (1996), which denotes repeated random sampling with replacement to produce a dataset of the same size as the original training set. The dataset is subsequently used for training of a base model and the collection of models obtained in this way forms an ensemble with individual models' decisions combined typically using voting (in classification problems) or averaging (in regression problems).

- *Boosting* proposed in Freund and Schapire (1997), which manipulates the probability with which samples are drawn from the original training data, to sequentially train classifiers focusing on the 'difficult' parts of the training set. Hence each consecutive ensemble member focuses on the training examples that cannot be successfully handled by the ensemble developed up to that point. The ensemble is usually built until a specified number of ensemble members is generated (although other stopping criteria are possible) and their decisions are combined using a weighted voting mechanism. Although the ensemble members can be 'weak' learners (i.e. models only slightly better than chance), this property must hold in the context of an increasingly difficult resampled dataset. As a result at some stage the 'weak' learner may in fact need to be quite complex and powerful.

The above approaches exploit variation in the data and are referred to as metalearning methods in Brazdil et al. (2009) and Vanschoren (2010). Bagging however does not satisfy point 2 of Definition 1, as consecutive random samples from the original dataset are independent from each other, so there is no experience from previous learning episodes involved. In the case of boosting however, the ensemble is built sequentially and it is the performance of previous ensemble members (i.e. experience gained while trying to solve the problem) that influences the sampling process.

More often, the following two approaches are considered as metalearning techniques:

- *Stacked generalisation* (or stacking) as introduced in Wolpert (1992), where a number of base learners is trained on the same dataset. Their outputs are subsequently being used for a higher level learning problem, building a model linking the outcomes of the base learners to the target value. The meta-model then produces the final target outcome.

- *Cascade generalisation* (Gama and Brazdil 2000), which works sequentially. When building a model, the output of the first base learner is appended to the original feature set and passed on to the next learner with the original target values. This process can then be repeated.

- Although in these cases the information about base-learning is drawn in the sense of point 2a of Definition 1, these algorithms are limited to a single problem domain with a bias that is fixed a priori, so that they, using the definition above, do not undoubtedly qualify as metalearning methods.

3.2 Algorithm recommendation

A considerable amount of metalearning research has been devoted to the area of algorithm recommendation. In this special case of metalearning, the aspect of interest is the relationship between data characteristics1and algorithm performance, with the final goal of predicting an algorithm or a set of algorithms suitable for a specific problem under study. As a motivation, the fact that it is infeasible to examine all possible alternatives of algorithms in a trial and error procedure is often given along with the experts necessary if pre-selection of algorithms is to take place. This application of metalearning can thus be both useful for providing a recommendation to an end-user or automatically selecting or weighting algorithms that are most promising.

Vanschoren (2010) points out another aspect: it is not only the algorithms themselves, but different parameter settings that will naturally let performance of the same algorithm vary on different datasets. It would be possible to regard versions of the same algorithm with different parameter settings as different learning algorithms altogether, but the author advocates treating the subject and studying its effects differently. Such an approach has for example been taken in Gomes et al. (2012) and Miranda et al. (2012), where the authors discuss a hybrid metalearning and search based technique to facilitate the choice of optimal parameter values of a Support Vector Machine (SVM). In this approach, the candidate parameter settings recommended by a metalearning algorithm are used a starting point for further optimization using Tabu Search or Particle Swarm Optimization techniques, with great success. Reif et al. (2012b) investigate increasing the accuracy and decreasing runtime of a genetic algorithm for selecting learning parameters for a Support Vector Machine and a Random Forests classifier. Based on past experience on other datasets and corresponding dataset characteristics, metalearning is used to select a promising initial population for the genetic algorithm, reducing the number of iterations needed to find accurate solutions.

An interesting treatment of the above problem can also be found in Jankowski and Grabczewski (2009), where the authors propose to take into account not only the expected performance of the algorithm but also its estimated training time. In this way the algorithms can be ordered according to the estimated training complexity, which allows to produce relatively well-performing models very quickly and then look for better solutions, while the ones already trained are producing predictions. These ideas are further extended in Jankowski (2011), where some modifications of the complexity measures used are introduced.

The classic application area of algorithm selection in machine learning is classification. Smith-Miles (2008) however tries to generalise the concepts to other areas including regression, sorting, constraint satisfaction and optimisation. Metalearning for algorithm selection has also been investigated in the area of time series forecasting, where the term was first used in Prudencio and Ludermir (2004b). A comprehensive and recent treatment of the subject can be found in Wang et al. (2009) and Lemke and Gabrys (2010), where time series are clustered according to their characteristics and recommendation rules or combination weights derived with machine learning algorithms. In the area of data mining, algorithm recommendation was identified as an important research issue at the 2001 KDD conference and the 2006 KDD workshops according to Brazdil et al. (2009).

Several systems for algorithm recommendation have been implemented. Following a previous successful European commission funded project, a project named 'meta-learning assistant for providing user support in machine learning and data mining' (METAL 2002), investigated model selection and combination approaches. A tool resulting from this project is the data mining advisor (DMA), a web-based system providing rankings of classification algorithms for users as described in Giraud-Carrier (2005), which however is no longer accessible. The last updates on the METAL project webpage date back to 2004/2005, and the webpage is currently offline. There is however another project focussing on preprocessing for data mining with a metalearning approach, called MiningMart (Morik and Scholz 2004). Although the software tool is still available for download,2 its most recent version is dated October 2006.

3.3 Dynamic bias selection

In classic algorithm recommendation, the bias depends on the available learning algorithm chosen and is not modified dynamically. Dynamic bias selection is often mentioned in relation to a continuous flow of training examples (data streams), where metalearning algorithms can be used for bias management and detecting concept drifts, as argued in Brazdil et al. (2009). In their metalearning review, Vilalta and Drissi (2002a) cite Bruha and Famili (2000) as the only example of dynamic bias selection apart from a few projects before 1992. In this contribution, a rule-based system which includes quality information about each rule, influencing the prediction process is used. The quality of a rule is updated continuously, which makes selection of the bias dynamic.

Very fast decision trees (VFDT) according to Domingos and Hulten (2000) dynamically adjust their biases with new incoming examples, with bias in this case referring to splitting tests in the tree nodes.

Another work qualifying as dynamic bias selection is the Learnt Topology Gating Artificial Neural Networks (LTGANN) by Kadlec and Gabrys (2008). In this contribution, neural networks are used for three different purposes: as base learners, as 'local experts' predicting the performance of one assigned base learner and as gating networks. The gating networks are used to discover relationships between different network topologies and their performance, influencing the topology of new networks added to the system.

3.4 Inductive transfer

A different flavour of metalearning runs under the name of 'inductive transfer', or 'learning to learn'. The emphasis here is not on investigating a preferably extensive set of problems as in algorithm selection; it is mostly used for a smaller selection of multiple related learning tasks. However, in the spirit of metalearning, knowledge from one problem domain is transferred across other domains or tasks. A learning mechanism thus accumulates experience that is supposed to improve performance through time. This notion puts a bigger focus on aspect 2 of Definition 1. Although aspect 1 still applies, most research in inductive transfer looks at a single learning mechanism and not a whole learning system.

Brazdil et al. (2009) distinguish two main forms of knowledge transfer using the example of neural networks. One is representational transfer, where training of the base learner is carried out sequentially. Functional transfer refers to training several neural networks in parallel while sharing all or a part of their internal structure. A special case of functional transfer is multitask learning, where a neural network uses output nodes to accommodate for more than one task. Evgeniou et al. (2005) discusses inductive transfer in kernel methods using the example of Support Vector Machines and regularisation networks. Single-task kernel learning algorithms are extended to multi-task learning by forcing all hypotheses to share a common component along with the problem-specific deviations. Kernel learning from a different perspective is presented in Aiolli (2012): a suitable kernel matrix for a linear Hebbian classifier is learnt with improving a basic kernel by learning chains of kernel transforms. Based on the results on an initial problem set, the kernel can then be applied to related problems.

In a special issue on metalearning, Silver and Bennett (2008) present several contributions in the field of inductive transfer. Zhang et al. (2008) approaches the subject from a Bayesian perspective, using a framework including latent variables for modelling a shared structure among different learning scenarios. Silver et al. (2008) uses a neural network with extra contextual inputs for different tasks. The task hence shifts to finding a single learner for the domains, only distinguished ('indexed') by the additional inputs.

In a recent survey article Pan and Yang (2010), inductive transfer is presented as a special case of a wider category of 'transfer learning', alongside transductive and unsupervised transfer learning. Transfer learning has been defined as a mechanism 'which aims to help improve the learning of the target predictive function in a new domain, using the knowledge gained in other domains'. Although this does not strictly adhere to Definition 1, the authors discuss relevant research issues of (1) what knowledge to transfer, (2) how to transfer it, and (3) in which situations to transfer the knowledge. This last point seems especially interesting as it can be equally well cast as in which situations not to transfer the knowledge, since it may happen that the source and destination domains are not related. In such a case the performance of the new model rather than benefiting, could suffer from the knowledge transfer (so called 'negative transfer') (Pan and Yang 2010).

3.5 Metalearning systems

The major usual steps of a modelling process consist of data analysis, data preprocessing, model building and a phase of interpretation and evaluation. Rather than applying metalearning in only a part of this process, latest research started taking a more holistic view by investigating metalearning frameworks and architectures.

While not yet mentioned in the seminal review of Vilalta and Drissi (2002a), Brazdil et al. (2009) give an overview of (semi-)automatic metalearning systems for data mining, only two of which target more than one aspect of the knowledge discovery process: Project CITRUS according to Wirth et al. (1997) seems to have been discontinued after a few publications in 1996 and 1997, however, it did target the complete modelling process. A bit more recently, Bernstein et al. (2005) propose the Intelligent Discovery Assistant (IDA), providing a template for ontology-driven assistants for knowledge discovery dealing with preprocessing, model building and post-processing. A realization of this idea has been materialized within the e-LICO project,3 which investigated the concept of 'meta-mining'—ontology and metaknowledge driven recommendation of data mining workflows as proposed in Nguyen et al. (2011) and evaluated in Nguyen et al. (2012). A somewhat related term, 'domain-driven data mining', also gained some recognition in the last years. The idea promotes a ubiquitous intelligence to be incorporated into the data mining process from a more methodological point of view, increasing reusability and understanding of algorithms. An overview of research done in this area and a discussion of future challenges and issues can be found in Cao (2010).

For continuous streams of data and the example of softsensors in process industry, Kadlec and Gabrys (2009) develop an architecture promoting life-long learning, where base-learners and pre- and post-processing methods can be plugged in and dynamically combined. A metalearning component optimises the system with regard to a global performance function. Jankowski and Grabczewski (2009) describe the implementation of an extensive metalearning architecture in detail, dealing with aspects like algorithm selection and parameter tuning, with the latter topic also addressed in Molina et al. (2012) for automatic tuning of parameters of decision trees.

METALA is an agent-based architecture with a metalearning component, which has been implemented as a J2EE based framework with details given in Hernansaez et al. (2004). It supports an arbitrary number of algorithms, which are managed by agents. Using metalearning on statistical and information-theoretic task properties, the best algorithm is dynamically selected, if the pool of algorithms and tasks is updated.

A distinct concept of lazy metalearning has been extensively studied in Bonissone (2012). The motivation for this work is the apparent lack of automation in model creation, which leads to bottlenecks in the models lifecycle and scalability (two other attempts at automation of model development methodology can also be found in Budka and Gabrys 2010; Budka et al. 2010). The main premise in lazy metalearning is that there is access to a large library of both local and global, pre-computed models together with their meta-information, which are perceived as commodity and which have been enabled by the recent expansion of cloud computing and crowdsourcing. The idea is to build dynamic on-demand ensembles using the models from the library as the building blocks, rather than optimizing and tuning of pre-computed models (Bonissone2012). Since exhaustive enumeration of all possible combinations of models in a library would be intractable, a multi-step selection process based on metaknowledge, query information and correlation of errors of the ensemble members is employed. According to Bonissone (2012), a multi-criteria decision making process is followed in order to (1) create the model by pre-selecting the initial building blocks for the assembly and compiling their meta-information, which is an off-line phase, and (2) perform dynamic model assembly, where the best subset of models for a given query is selected on-line, i.e. during runtime. The process is also able to determine the weights of the ensemble models in the fusion schema, based on their local performance around the query. Some other recent approaches to development of metalearning frameworks can be found in Matijaš et al.

(2013) for the problem of electricity load forecasting, Abbasi et al. (2012) for financial fraud detection or Tsai and Hsu (2013) for bankruptcy prediction.

4. CONSIDERATIONS FOR USING METALEARNING

Before applying metalearning to any problem, certain practical choices have to be made. This includes the choice of a metalearning algorithm, which can even constitute a meta-metalearning problem itself. Selection of appropriate metaknowledge and the problem of setting up and maintaining metadatabases have to be tackled, research efforts of which will be summarised in this section.

4.1 Prerequisites

As also elaborated on in Brazdil et al. (2009), metalearning can not be seen as a magic cure to machine learning problems for a variety of reasons. First of all, the extracted metafeatures need to be representative of their problem domain, otherwise, an algorithm will fail to identify similar domains. On the same note, if a problem has not been seen before, metalearning will be unable to exploit past knowledge to improve prediction performance. Performance estimation may be unreliable because of the natural limitations of estimating the true performance of the dataset. Different metafeatures might be applicable to each dataset. These issues emphasise the importance of being critical when designing a metalearning system.

4.2 Metalearning algorithms

Vanschoren (2010) gives a survey on efforts to describe properties of algorithms. The author distinguishes qualitative properties (for example type of data that can be handled, learning strategy, incrementality) and quantitative properties (bias-variance profile, runtime properties like scalability and resilience). In an effort to find an implementation and vendor-independent method for representing machine learning models, the XML-based standard PMML has been developed and gained some recognition in the last years. A detailed description of PMML can be found in Guazzelli et al. (2009).

The choice of a metalearning algorithm naturally depends on the problem and the task to be solved. Generally, traditional classification algorithms are very successful in metalearning algorithm selection and can include meta-decision trees (Todorovski and Dzeroski 2003), neural networks, Support Vector Machines or any other classification algorithms, with the *k*-Nearest Neighbours being another popular choice (Brazdil et al. 2009). Applying regression algorithms is less popular, even smaller is the number of available algorithms to learn rankings. One of the simplest ranking method involves dividing the problem space using clustering of available datasets according to a distance measure (usually *k*-Nearest Neighbour) of the metafeatures and using average performance ranks of the cluster into which a new problem falls (Brazdil et al. 2003). Brazdil and Soares (2000) also look at the magnitude and significance of the differences in performance. The NOEMON approach introduced by Kalousis and Theoharis (1999) builds classifiers for each pair of base forecasting methods with a ranking being generated using the classifiers' outputs. Todorovski et al. (2002) build decision trees using the positions in a ranking as target values.

4.3 Extracting metaknowledge

According to Brazdil et al. (2009), metaknowledge is derived in the course of employing a learning system. A very common form of metaknowledge is the performance of algorithms in certain problem domains, which is to be linked with characteristics of the task. Several possibilities for characterising a problem domain exist.

The most straightforward form of metaknowledge extracted from the data include statistical or information-theoretic features. For classification problems, Brazdil et al. (2009) mention the number of classes and features, ratio of examples to features, degree of correlation between features and target concept and average class entropy. For other application areas, features can look completely different, as for example summarised in Lemke et al. (2009) for the area of time series forecasting, where features can include, for example, length, seasonality, autocorrelation, standard deviation and trends of the series.

Vilalta and Drissi (2002b) propose measures for the difficulty of a classification problem that can be used as an input for metalearning. They include class variation, denoting the probability that, by means of a distance measure, any two neighbouring data records have a different class value and example cohesiveness, measuring the density of the example distribution in the training set. In a similar approach, Köpf and Iglezakis (2002) also suggest comparing observations with each other and extract 'case base properties', which assess the quality of a dataset using measures such as redundancy, for example induced by data records that are exactly the same, or incoherency, which, for example occurs if data records have the same features but different class labels.

Alternatively to looking at the data only, information of individual algorithms and how they solved the problem can be considered, for example their predicted confidence intervals. This can be achieved by using a model that is fast to build and train and investigating its properties. In this spirit, Bensusan et al. (2000) suggest building a decision tree for a classification problem and using properties of the tree such as nodes per feature, tree depth or shape to characterise it. Another approach is landmarking as proposed in Pfahringer et al. (2000), using the performance of simple algorithms to describe a problem and correlating this information with the performance of more advanced learning algorithms. A list of landmarking algorithms can be found in Vanschoren (2010). Landmarking algorithms can also be run on only a small sample of the data available, reducing the training time required. Performance information of different algorithms and learning curves generated when more data is added to the training set can then be used to select an algorithm according to Fürnkranz et al. (2002).

Empirical evaluation of different categories of metafeatures in the context of their suitability for predicting classification accuracies of a number of standard classifiers can be found in Reif et al. (2012c). The authors distinguish 5 such categories of features i.e. simple, statistical, information-theoretic, landmarking and model-based, which corresponds to the general categorization evident from the literature.

As with any learning problem, metalearning is subject to the 'curse of dimensionality' (Bishop 1995) and other issues, which can traditionally be solved by selecting a subset of relevant features. Although to the best of our knowledge, in the context of metalearning this issue has only been addressed in relatively few publications (e.g. Kalousis and Hilario 2001; Reif et al. 2012c; Todorovski et al. 2000), we assume that the reason for this is quite simple—meta-feature selection does not differ from feature selection at the base-level, and the machine learning literature is very rich in this regard (a comprehensive review of various feature selection techniques can be found in Guyon and Elisseeff 2003).

4.4 Metadatabases

As metalearning profits from knowledge obtained while looking at data from other problem domains, having sufficient datasets at one's disposal is important. Soares (2009) propose transforming existing datasets ('datasetoids') to obtain a larger number of them and show success of the approach on a metalearning post-processing problem. Vanschoren (2010) states that there is no lack of experiments being done, but datasets and information obtained often remain in 'people's heads and labs'. He proposes a framework to export experiments to specifically designed experiment databases based on an ontology for experimentation in machine learning. The resulting database can then, for example, give information on rankings of learning algorithms, the behaviour of ensemble methods, learning curve analyses and the bias-variance behaviour of algorithms. One example of such database can be The Open Experiment Database.4 An analysis of this database together with a critical review can be found in Driessens et al. (2012).

An alternative approach to the problem of scarcity metadatabases has been presented in Reif et al. (2012a), where the authors describe a dataset generator able to produce synthetic datasets with specified values of some metafeatures (like kurtosis and skewness). Although the proposed generator appears to be at a very early stage of development, the idea is definitely very promising, also from the point of view of performing controlled experiments on datasets with specified properties. Similarly to feature selection, synthetic data generation has received a considerable attention in the recent generic machine learning and data mining literature, especially in the context of data streams and concept drift (please see Bifet et al. 2011 and references therein).

5. CONCLUSIONS AND RESEARCH CHALLENGES

Research in the area of metalearning is continuing in several directions. One area is the identification of metafeatures. As mentioned before, the vast majority of publications investigates extracting features from the dataset, mostly in the form of statistical or information theoretic measures. Landmarking is a different approach using simple base learning algorithms and their performance to describe the dataset at hand. However, Brazdil et al. (2009) argue that characteristics of learning algorithms and gaining a better understanding of their behaviour would be a valuable research avenue with very few publications, for example Vanschoren and Blockeel (2006), that exist in this area to date.

A lot of publications on metalearning focus on selecting the base-learning method that is most likely to perform well for a specific problem. Fewer publications like Brazdil et al. (2003) and Prudencio and Ludermir (2004a) consider ranking algorithms, which can be used to guide combination weights and to increase robustness of a metalearning system.

Regarding adaptivity and continuous monitoring, many approaches go further than the static traditional metalearning approaches, for example by using architectures that support life-long learning such as in Kadlec and Gabrys (2009). However, research in this area can still go a long way further investigating continuous adjustment, rebuilding or discarding of base-learners with the help of metalearning approaches.

Users of predictive systems are faced with a difficult choice of an ever increasing number of models and techniques. Metalearning can help to reduce the amount of experimentation by providing dynamic advice in form of assistants, decrease the time that has to be spent on introducing, tuning and maintaining models and help to promote machine learning outside of an academic environment. In this context, architectures and frameworks using metalearning have been named in Sect. 3.2. However, many of these are work in progress, no longer maintained or tackle only one of the aspects in a modelling process, so this area would benefit from further research and implementations as well.

FOOTNOTES

1 In a more advanced form it is the relationship between data and algorithm characteristics, and algorithm performance.

2 http://www-ai.cs.uni-dortmund.de/MMWEB/downloads.

3 http://www.e-lico.eu/.

4 http://expdb.cs.kuleuven.be/.

ACKNOWLEDGMENTS

The research leading to these results has received funding from the European Union 7th Framework Programme (FP7/2007-2013) under grant agreement no. 251617.

REFERENCES

1. Abbasi A, Albrecht C, Vance AO, Hansen JV (2012) Metafraud: a meta-learning framework for detecting financial fraud. Manag Inf Syst Q 36(4):1293–1327

2. Aiolli F (2012) Transfer learning by kernel meta-learning. J Mach Learn Res Proc Trac 27:81–95

3. Bensusan H, Giraud-Carrier C, Kennedy C (2000) A higher-order approach to meta-learning. In: Proceedings of the ECML'2000 workshop on meta-learing: building automatic advice strategies for model selection and method combination

4. Bernstein A, Provost F, Hill S (2005) Toward intelligent assistance for a data mining process: an ontology-based approach for cost-sensitive classification. IEEE Trans Knowl Data Eng 17:503–518

5. Bifet A, Holmes G, Kirkby R, Pfahringer B (2011) Data stream mining a practical approach. Technical report. The Unibversity of Waikato

6. Biggs JB (1985) The role of meta-learning in study process. Br J Educ Psychol 55:185–212

7. Bishop C (1995) Neural networks for pattern recognition. Oxford University Press, New York

8. Bonissone PP (2012) Lazy meta-learning: creating customized model ensembles on demand. In: Advances in computational intelligence, Springer, Heidelberg pp 1–23

9. Brazdil P, Soares C, de Costa P (2003) Ranking learning algorithms: using IBL and meta-learning on accuracy and time results. Mach Learn 50(3):251–277

10. Brazdil P, Giraud-Carrier C, Soares C, Vilalta R (2009) Metalearning: applications to data mining. Springer, Berlin

11. Brazdil P, Soares C (2000) A comparison of ranking methods for classification algorithm selection. In: de Mantaras R, Plaza E (eds) Machine learning: proceedings of the 11th European conference on machine learning ECML2000, pp 63–74. Springer

12. Breiman L (1996) Bagging predictors. Mach Learn 24(2):123–140

13. Bruha I, Famili A (2000) Postprocessing in machine learning and data mining. ACM SIGKDD Explor Newslett 2:110–114

14. Budka M, Gabrys B (2010) Ridge regression ensemble for toxicity prediction. Procedia Computer Science 1(1):193–201.

15. Budka M, Gabrys B, Ravagnan E (2010) Robust predictive modelling of water pollution using biomarker data. Water Res 44(10):3294–3308.

16. Cao L (2010) Domain-driven data mining: challenges and prospects. IEEE Trans Knowle Data Eng 22:755–769

17. de Miranda P, Prudêncio R, de Carvalho A, Soares C (2012) An experimental study of the combination of meta-learning with particle swarm algorithms for svm parameter selection. In: Computational science and its applications-ICCSA, pp 562–575

18. Dietterich T (2000) Ensemble methods in machine learning. In: Proceedings of the first international workshop on multiple classifier systems, pp 1–15

19. Domingos P, Hulten G (2000) Mining high-speed data streams. In: Proceedings of the sixth ACM SIGKDD international conference on knowledge discovery and data mining, pp 71–80

20. Driessens K, Vanwinckelen G, Blockeel H (2012) Meta-learning from an experiment database. Proceedings of the Workshop on Teaching Machine Learning at the 29th International Conference on Machine Learning. Edinburgh, UK

21. Evgeniou T, Micchelli C, Pontil M (2005) Learning multiple tasks with kernel methods. J Mach Learn Res 6:615–637

22. Freund Y, Schapire RE (1997) A decision-theoretic generalization of on-line learning and an application to boosting. J Comput Syst Sci 55(1):119–139.

23. Fürnkranz J, Petrak J, Brazdil P, Soares C (2002) On the use of fast subsampling estimates for algorithm recommendation. Technical report. sterreichisches Forschungsinstitut fr Artificial Intelligence

24. Gama J, Brazdil P (2000) Cascade generalisation. Mach Learn 41(3):315–343

25. Giraud-Carrier C (2005) The data mining advisor: meta-learning at the service of practitioners. In: Proceedings of the fourth international conference on machine learning and applications, ICMLA '05, pp 113–119. IEEE Computer Society, Washington, DC, USA

26. Giraud-Carrier C (2008) Metalearning—a tutorial. In: Tutorial at the 7th international conference on machine learning and applications (ICMLA), San Diego, California, USA

27. Gomes TA, Prudêncio RB, Soares C, Rossi AL, Carvalho A (2012) Combining meta-learning and search techniques to select parameters for support vector machines. Neurocomputing 75(1):3–13

28. Guazzelli A, Zeller M, Lin WC, Williams G (2009) PMML: an open standard for sharing models. R J 1(1): 60–65

29. Guyon I, Elisseeff A (2003) An introduction to variable and feature selection. J Mach Learn Res 3:1157–1182

30. Hernansaez JM, Botia JA, Skarmeta AF (2004) METALA: a J2EE technology based framework for web mining. Revista Colombiana de Computación 5(1)

31. Jankowski N (2011) Complexity measures for meta-learning and their optimality. In: Solomonoff 85th memorial. Lecture notes in computer science. Springer

32. Jankowski N, Grabczewski K (2009) Universal meta-learning architecture and algorithms. In: Duch W, Grabczewski K, Jankowski N (eds) Meta-learning in computational intelligence. Springer, Berlin

33. Kadlec P, Gabrys B (2009) Architecture for development of adaptive on-line prediction models. Memet Comput 4(1):241–269

34. Kadlec P, Gabrys B (2008) Learnt topology gating artificial neural networks. In: Proceedings of the international joint conference on neural networks (IJCNN 2008) as part of the 2008 IEEE World Congress on Computational Intelligence (WCCI'2008), pp 2605–2612

35. Kalousis A, Theoharis T (1999) NOEMON: design, implementaion and performance results of an intelligent assistant for classifier selection. Intell Data Anal 5(3):319–337

36. Kalousis A, Hilario M (2001) Feature selection for meta-learning. In: Cheung D, Williams G, Li Q (eds) Advances in knowledge discovery and data mining, pp 222–233. Springer

37. Köpf C, Iglezakis I (2002) Combination of task description strategies and case base properties for meta-learning. In: Proceedings of the 2nd international workshop on integration and collaboration aspects of data mining, decision support and meta-learning, pp 65–76

38. Lemke C, Gabrys B (2010) Meta-learning for time series forecasting and forecast combination. Neurocomputing 73(10):2006–2016

39. Lemke C, Riedel S, Gabrys B (2009) Dynamic combination of forecasts generated by diversification procedures applied to forecasting of airline cancellations. In: Proceedings of the IEEE symposium series on computational intelligence, pp 85–91

40. Matijaš M, Suykens JA, Krajcar S (2013) Load forecasting using a multivariate meta-learning system. Expert Syst Appl 40(11):4427–4437

41. Metal: meta-learning assistant for providing user support in machine learning and data mining. http://www.metal-kdd.org/ (2002)

42. Molina MDM, Romero C, Ventura S, Luna JM (2012) Meta-learning approach for automatic parameter tuning: a case study with educational datasets. In EDM, pp 180–183

43. Morik K, Scholz M (2004) The miningmart approach to knowledge discovery in databases. In: Intelligent technologies for information analysis. Springer, Heidelberg, pp 47–65

44. Nguyen P, Kalousis A, Hilario M (2011) A meta-mining infrastructure to support kd workflow optimization. In: eCML PKDD 2011, p 1

45. Nguyen P, Kalousis A, Hilario M (2012) Experimental evaluation of the e-lico meta-miner. In: 5th planning to learn workshop WS28 at ECAI 2012, p 18

46. Pan S, Yang Q (2010) A survey on transfer learning. IEEE Trans Knowl Data Eng 22(10):1345–1359

47. Pfahringer B, Bensusan H, Giraud-Carrier C (2000) Meta-learning by landmarking various learning algorithms. In: Proceedings of the seventeenth international conference on machine learning. Morgan Kaufmann, pp 743–750

48. Prudencio RB, Ludermir TB (2004a) Meta-learning approaches to selecting time series models. Neurocomputing 61:121–137

49. Prudencio R, Ludermir T (2004b) Using machine learning techniques to combine forecasting methods. In: Proceedings of the 17th Australian joint conference on artificial intelligence, pp 1122–1127

50. Reif M, Shafait F, Dengel A (2012a) Dataset generation for meta-learning. In KI-2012: Poster and Demo Track, pp 69–73

51. Reif M, Shafait F, Dengel A (2012b) Meta-learning for evolutionary parameter optimization of classifiers. Mach Learn 87:357–380.

52. Reif M, Shafait F, Goldstein M, Breuel T, Dengel A (2012c) Automatic classifier selection for non-experts. Pattern Anal Appl 1–14.

53. Rice J (1976) The algorithm selection problem. In: Rubinov M, Yovits MC (eds) Advances in computers, vol 15. Academic Press, Inc, London

54. Silver D, Bennett K (2008) Guest editor's introduction: special issue on inductive transfer learning. Mach Learn 73:215–220

55. Silver D, Poirier R, Currie D (2008) Inductive transfer with context-sensitive neural networks. Mach Learn 73(3):313–336

56. Smith-Miles KA (2008) Cross-disciplinary perspectives on meta-learning for algorithm selection. ACM Comput Surv 41(1):1–25

57. Soares C (2009) Uci++: improved support for algorithm selection using datasetoids. In: Theeramunkong T, Kijsirikul B, Cercone N, Ho T-B (eds) Advances in knowledge discovery and data mining. Springer, Berlin

58. Todorovski L, Blockeel H, Dzeroski S (2002) Ranking with predictive clustering trees. In: Elomaa T, Mannila H, Toivonen H (eds) Proceedings of the 13th European conference on machine learning. Springer, pp 444–455

59. Todorovski L, Brazdil P, Soares C (2000) Report on the experiments with feature selection in meta-level learning. In: Proceedings of the PKDD-00 workshop on data mining, decision support, meta-learning and ILP: Forum for Practical Problem Presentation and Prospective Solutions. Citeseer

60. Todorovski L, Džeroski S (2003) Combining classifiers with meta decision trees. Mach Learn 50(3):223–249

61. Tsai CF, Hsu YF (2013) A meta-learning framework for bankruptcy prediction. J Forecast 32(2):167–179

62. Vanschoren J (2010) Understanding machine learning performance with experiment databases. PhD thesis, Arenberg Doctoral School of Science, Engineering & Technology, Katholieke Universiteit Leuven

63. Vanschoren J, Blockeel H (2006) Towards understanding learning behavior. In: Proceedings of the annual machine learning conference of Belgium and the Netherlands, pp 89–96

64. Vilalta R, Drissi Y (2002a) A perspective view and survey of meta-learning. Artif Intell Rev 18:77–95

65. Vilalta R, Drissi Y (2002b) A characterization of difficult problems in classification. In: Proceedings of the 6th European conference on principles and practice of knowledge discovery in databases, Helsinki, Finland

66. Wang X, Smith-Miles K, Hyndman R (2009) Rule induction for forecasting method selection: meta-learning the characteristics of univariate time series. Neurocomputing 72:2581–2594

67. Wirth R, Shearer C, Grimmer U, Reinartz T, Schloesser J, Breitner C, Engels R, Lindner G (1997) Towards process-oriented tool support for kdd. In: Proceedings of the 1st European symposium on principles of data mining and knowledge discovery, Trondheim, Norway

68. Wolpert D (1992) Stacked generalization. Neural Netw 5:241–259

69. Yao X, Islam M (2008) Evolving artificial neural network ensembles. IEEE Comput Intell Mag 3:31–42

70. Zhang J, Ghahramani Z, Yang Y (2008) Flexible latent variable models for multi-task learning. Mach Learn 73(3):221–242

Chapter 2

Automatic Multi-Document Arabic Text Summarization Using Clustering and Keyphrase Extraction

Hamzah Noori Fejer and Nazlia Omar

Center for AI Technology, Faculty of Information Science and Technology, Universiti Kebangsaan Malaysia (UKM), Bangi, Selangor, Malaysia

ABSTRACT

Automatic text summarization has become important due to the rapid growth of information texts since it is very difficult for human beings to manually summarize large documents of texts. A full understanding of the document is essential to form an ideal summary. However, achieving full understanding is either difficult or impossible for computers. Therefore, selecting important sentences from the original text and presenting these sentences as a summary present the most common techniques in automated text summarization. Arabic natural language processing lacks tools and resources which are essential to advance research in Arabic text summarization. In addition to the limited resources, there has been little attention and research done in this field. Arabic text summarization still suffer from low accuracy as they use simple summarization techniques. The aim of this research is to improve Arabic text summarization by using clustering and keyphrase extraction. This study proposes a combined clustering method to group Arabic documents into several clusters. Keyphrase extraction module is applied to extract important keyphrases from each cluster, which helps to identify the most important sentences and find similar sentences based on several similarity algorithms. These algorithms are applied to extract one sentence from a group of similar sentences while ignoring the other similar sentences. The Recall-Oriented Understudy for Gisting Evaluation (ROGUE) metrics were used for the evaluation. For the summarization dataset the corpus DUC2002 was used. This model achieved an accuracy of 43.4%. The experiments have proved that the proposed model has given better performance in comparison to other work.

1. INTRODUCTION

With summarization, one would simply summarize one article or document by creating a summary of the concepts that are most important. The summary would be in the correct order and coherent as well. This process takes a lot of effort and is time consuming, even for humans. A summary of the same article that is created by two different people would be different, based on what each person thinks is most important and includes this in their summary; they each have a unique perspective on the same article. This encouraged the requirement of an automated summarizer system that may carry out this process with less effort and time. This has created a large amount of research on this topic of automatic summarization, which was over 50 years ago (Luhn, 1958).

Generally, summarization of text is the task of creating a summary from an article or a collection of articles about the same subject by extracting the most important parts of the text and adding them in chronological order. However, automatic text summarization is the main task of generating a shorter version of a particular document or set of documents by machines. For instance, a document of text may be compressed version via the usage of applications.

A summary in general must provide the core concepts within the text that is input. Only main sentences must be available in the summary and the task of explaining the sentences depends on the method that is implemented for summarization. With the task of automatic summarization, there are generally two primary methods that are widely carried out. These approaches are extractive and abstractive. In the first approach, which is extractive based summarization, there is a limit in the extraction and only the important sentences are extracted and generated into a summary in the chronological order, in order for a coherent summary to be generated. The parts of text that are extracted vary, based on the summarizer used.

Numerous summarizers take sentences, as opposed to paragraphs, or other larger text units. Automatic text summarization is carried out often using extractive based approaches. However, abstractive based summarization deals with more tools that rely on language and Natural Language Generation (NLG) concepts. This type of summarization may incorporate terms that do not exist in the article. Abstractive summarization aims to copy methods used by humans, such as adding a concept that is available in the original article in a better and more comprehensive way. This type of summarization, however, is more complex to implement, although more effective. This work uses an extractive concept implemented on the proposed model.

Summarization has been investigated for over the past 50 years, with most studies focusing on the English language (Hirao *et al.*, 2007). Studies on Arabic-language documents have been conducted at a much later period and remains far behind those studies on other languages. Research on automatic summarization of Arabic-language documents has started approximately 10 years ago (Conroy *et al.*, 2006; Douzidia and Lapalme, 2004). Further studies on Arabic-language resources are required. Several studies have proposed relatively advanced methods for Arabic language summarization (El-Haj *et al.*, 2011a, b; Giannakopoulos *et al.*, 2008).

The task of clustering within summarization of documents may be critical for choosing and retrieving related sentences and removing any redundancies. Even though approaches that include clustering have become widely used in text summarization, there has been a lack of research for clustering methods in summarization of documents in the Arabic language. In this work, focus is made on hierarchical and partitional clustering methods, namely, single linkage, complete linkage and k-means clustering because it is suitable with cluster summarization (Liu and Croft, 2004). This study propose a cluster-based summarization approach to group similar texts as well as employs a keyphrase extraction technique by an unsupervised machine learning algorithm to recognize sentences that include keyphrases and to summarize original text documents.

2. MATERIALS AND METHODS

This section describes the proposed framework for multi-document Arabic text summarization. A pre-processing text, clustering algorithms has been applied to group documents into many clusters. Keyphrase extraction is used to extract the important Keyphrases from each cluster, which helps to identify the most important sentences. Eliminate redundancy technique is applied to extract one sentence from a group of similar sentences and ignoring the other. Figure 1 shows the several phases in the overall architecture of our system, which are outlined as follows.

Input documents: Input multi-documents will be used to extract texts from articles on different subjects, such as education, politics and sports. These documents contain texts of different sizes. The C# language is used to code the different stages of the proposed summarization model.

Text pre-processing: This phase involve four steps, namely, tokenizing, eliminating stop words, stemming and text representation and term weighting.

Clustering phase: Clustering is classifying data into many clusters. The elements within a cluster are similar to one another, but are dissimilar to elements in another cluster. Clustering decreases the amount of information by categorizing and grouping similar data. Document clustering involves unsupervised document organization, automatic topic extraction and fast **information retrieval**. Clustering is performed to group these results automatically into a list of meaningful categories.

Hierarchical clustering is the most commonly used clustering method. Hierarchical clustering method constructs the clusters by recursively partitioning the instances in either a top-down or bottom-up fashion. Single linkage clustering and Complete linkage clustering are used in this study.

Single linkage clustering (Nearest Neighbor Technique) is one of the simplest agglomerative clustering techniques. The distance between two groups is defined as that of the closest pair of individuals. One individual is considered from each group. Single linkage clustering is considered the best hierarchical method in terms of theoretical soundness. Several efficient methods are devised to apply the single linkage method on large data sets.

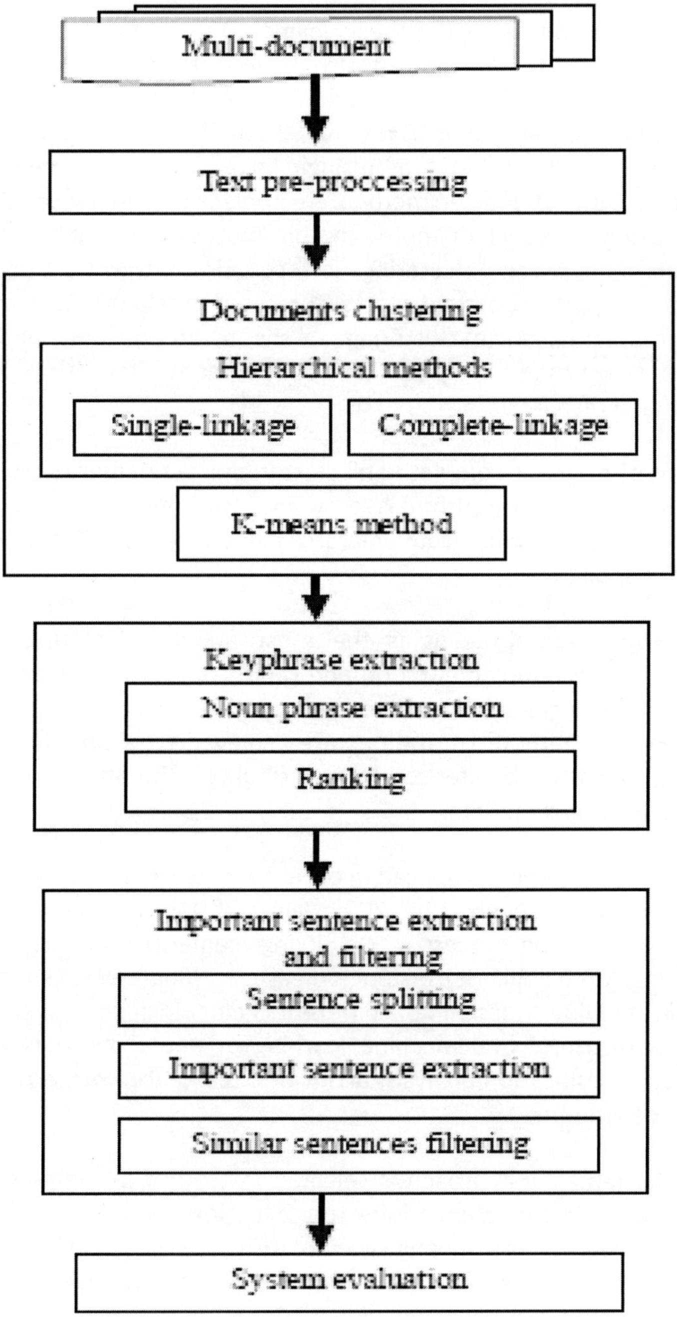

Figure 1. Overall architecture of the multi-document arabic text summarization

Complete linkage clustering (Furthest Neighbor Clustering) is the reverse of single linkage clustering in the sense that the distance between groups is defined as the distance between the most distant pair of individuals that are selected from each group. The definition of cluster membership in this method is much stricter than that in single linkage clustering and the large, straggly clusters in the latter method are replaced by many small, tightly bound clusters of equal diameter in this method. In graph theoretical terms, complete linkage clustering corresponds to the identification of maximally complete sub-graphs at a threshold similarity. Willett (1982) defined complete linkage clustering as the most effective method for document clustering despite its requirement for the greatest computational resources.

The output of single-or complete-link clustering is a number of clusters (K-value). The k-means algorithm is easy to apply, because a number of clusters are determined from hierarchical clustering. Then, k-means clustering is applied.

K-means clustering: K-means is the most famous clustering method, because of its easy implementation and rapid convergence. Nevertheless, this method is significantly influenced by the selection of the initial solution. In this method, t iterations are achieved on a sample size of m items that are each characterized by n attributes. t: iteration k: number of clusters n: attributes m: item.

A lot of works have been proposed in order to improve the efficiency of the original k-means (Kanungo *et al.*, 2002; Jain and Dubes, 1988) defines that **cluster analysis** is the process of classifying elements into sets of similar objects based on a similarity/distance measure. Nonetheless, k-means are applied in a wide number of different fields including text mining, **information retrieval** and machine learning of **neural network**, **pattern recognition**, classification analysis, **artificial intelligence**, **image processing** and machine vision.

As such, the k-means approach is selected instead of its fast variants for comparative purposes throughout the experiments conducted in this research. Moreover, as k-means is adopted in the algorithm proposed in this study, fast variants of k-means can be used to improve its speed K-means is a method that has been widely used for partitional clustering with a linear time complexity (Cimiano *et al.*, 2005). As stated by Hartigan (1975), the k-means algorithm argues that the mean of the documents assigned to that cluster represents. each of the k-clusters and as a result the k-means technique is largely regarded as the centroid of that cluster. The benefit of k-means clustering are simple and flexible easy to understand and can be easily to implemented.

The disadvantage of k-means clustering method is clusters number (k-value) must be determined at the beginning, but after applying k-means algorithm, number of clusters may not be the same as the value specified in advance (Vora and Oza, 2013). For this reason we proposed consolidation method combining hierarchical clustering to get the exact number of clusters (topics) and k-means clustering. Then, keyphrases are extracted from each cluster (each topic will be summarized, separately).

Keyphrase extraction: Keyphrases are important words/phrases that reflect the subject of the text. We extract keyphrases from each cluster during this phase. Keyphrase extraction comprises the following steps:

Noun phrase extraction: In this phase, all possible noun phrases of one, two, three, or four consecutive words that appear in a given document are generated as n-gram terms. Keyphrase often corresponds to frequently occurring noun phrase in a text. This extraction algorithm considers only noun phrases as candidate keyphrases.

Ranking: For each noun phrase, some set of features are extracted. The following is used for ranking the candidate keyphrase:

- Sentence count: This feature refers to the total number of sentences in which members of the keyphrase can be found. We normalize this feature using the total number of sentences in the document

- Term frequency: Frequency refers to the number of occurrences of the candidate phrase. Frequency is calculated using an aggressive stemming algorithm (iterated Light stemmer). Frequency is normalized by the number of noun occurrences in the document

- First occurrence in text: This feature refers to the first occurrence of a word in the text. This feature is normalized by the total number of sentences in the document

- Last occurrence in text: This feature refers to the last occurrence of a word in the text. This feature is normalized by the total number of sentences in the document

- C-value: This feature is normalized for potential keyphrases by using the length of the phrase and the frequency, with which it occurs as a sub-string of other phrases

Important sentences extraction and filtering: This phase extracts the most important sentences from each cluster. This phase involves the following steps:

Sentence splitting: In this step, each document is splitting into several sentences using delimiters (e.g., full stop, question mark and exclamation mark).

Important sentences extraction: Obviously, every sentence acquired from a text is not important for text summarization.

We assume that sentences containing keyphrases are only important. Therefore, sentences set can be filtered to provide better results by removing sentences without contain any keyphrases. The score of sentence yields:

$$\text{Score (sentence)} = \frac{\text{Total \# of words in every keyphrases in the sentence}}{\text{Sentencs length in words}} \tag{1}$$

Furthermore, it is assumed that the sentence is considered as important, if the is greater than a threshold otherwise it is excluded from the sentences set.

Redundancy elimination: In this phase, all sentences are taken in each cluster. Similarity measures are used to identify similar sentences. Then, similar sentences are coherently grouped. The well-known measures of distance between patterns are majorly focused in this study. Consequently, the Cosine Similarity (CS) and Jaccard Coefficient (JC) are computed as similarity or distance measures (Cimiano *et al.*, 2005; Huang, 2008). Finally, one sentence is selected by ignoring others. The selected sentence becomes the most important sentence based on the calculated score. The CS and JC are defined as follows:

Cosine similarity: Cosine similarity is one of the most famous similarity measures. The cosine similarity of two sentences, namely, $\vec{t_a}$ and $\vec{t_b}$ is computed as follows:

$$SIM_C\left(\vec{t_a}, \vec{t_b}\right) = \frac{\vec{t_a} \cdot \vec{t_b}}{|\vec{t_a}| * |\vec{t_b}|} \tag{2}$$

where, $\vec{t_a}$ and $\vec{t_b}$ are the m-dimensional vectors over the term set T = {t1, tm}. Each dimension represents the positive weight of a term in the document.

Jaccard coefficient: The Jaccard coefficient, also known as the Tanimoto coefficient, is used to measure the similarity in the intersection that is divided by the union of the objects. For text sentences, the Jaccard coefficient compares the sum weight of shared terms and the sum weight of non-shared terms that are presented in either of the two sentences. The Jaccard coefficient is formally defined as follows:

$$SIM_j\left(\vec{t_a}, \vec{t_b}\right) = \frac{\vec{t_a} \cdot \vec{t_b}}{|\vec{t_a}|^2 + |\vec{t_b}|^2 - \vec{t_a} \cdot \vec{t_b}} \tag{3}$$

The Jaccard coefficient equals to 1 when $\vec{t_a} = \vec{t_b}$ and equals to 0 when $\vec{t_a}$ and $\vec{t_b}$ are disjointed. The corresponding distance measure is computed as $D_j = 1\text{-}SIM_j$.

Evaluation: Evaluating the accuracy of a generated summary is difficult, because an optimal summary remains undefined (Fiszman *et al.*, 2009). Achieving system evaluation may help address this problem.

In our work, ROUGE-N is used to calculate the scores of a candidate summary based on the n-gram overlap between candidate and reference summaries (Lin, 2004). ROUGE-N consists of many metrics, such as ROUGE-1, ROUGE-2, ROUGE-3 and so on, with each corresponding to the size of the n-grams used in the evaluation. In this study N is 1-4. ROUGE-N scores are computed as follows:

Thus, the recall (R) measure calculates the proportion of n-grams from reference summaries that occur in a candidate summary, the precision (P) calculates the proportion of n-grams from a candidate summary that occur in reference summaries and F-score combines recall and precision into one metric. The following formulas compute, recall, precision and F-score, as:

$$Recall - N = \frac{|G_{ref} \cap G_{can}|}{|G_{ref}|} \quad (4)$$

$$Precision - N = \frac{|G_{ref} \cap G_{can}|}{|G_{can}|} \quad (5)$$

$$ROUGE - NF - score = \frac{2 * ROUGE - Nrecall * ROUGE - Nprecision}{ROUGE - Nrecall + ROUGE - Nprecision} \quad (6)$$

where, G_{ref} includes the grams of reference summary and G_{can} includes the grams of candidate summary.

3. RESULTS AND DISCUSSION

Data description: For the document collection, machine translation was used to automatically create an Arabic multi-document summary corpus. The machine translation was used to translate an English data set into Arabic. The output of the translation process is an Arabic data set of related news articles. The data set used in the machine translation process was the DUC-2002 data set provided freely by the National Institute of Standards and Technology (NIST) through DUC2002. DUC-2002 is an English data set that contains 567 articles, 17,340 sentence and 199,423 word in addition to 1,111 gold-standard (model) summaries. National Institute of Standards and Technology produced 60 reference sets.

Experimental results: Five clustering methods were examined: k-means with cosine algorithm (KC), k-means with Jaccard coefficient algorithm (KJ), Single Linkage (SL), Complete Linkage (CL) and combined Clustering Method (CM). A keyphrase extraction module was then applied to extract important keyphrases from each cluster, identify the most important sentences and find similar sentences based on several similarity algorithms to eliminate redundancy. For test this model, three different sets of experiments were carried out in this study. The first experiment aimed to summarize three documents that are classified under the educational category, the second experiment was conducted to summarize five documents that belong to the political category and the last experiment was carried out to summarize eight documents that are classified under the two categories (educational and political). Table 1 shows the results of these experiments, respectively.

All experiments were designed with the number of clusters (k value). This number was specified by the user as two and is randomly initiated in the k-means methods. The number of clusters in the hierarchical methods is based on the distance/similarity between the documents was between 0.03-0.05. The threshold value chosen for the experiments was 0.05. The ROUGE-1 (n-gram = 1) were considered for the evaluation of this study. According to these results, the clustering by using a Combined Method (CM) was better and more effective than the k-means methods and the hierarchical methods. Based on the results, the best overall recall, precision and ROUGE-1, which were achieved in the second experiment. Figure 2 shows the compression of this model's summaries with the summaries of El-Haj (2012), which is the most recent related work. The recall, precision and ROUGE-1 scores of the El-Haj system were 39.52, 38.49 and 38.99%, respectively, whereas those of this model were 45.23, 41.80 and 43.45%, respectively, as obtained from the second experiment. The experiments proved that this model achieved better results than the El-Haj system.

4. CONCLUSION

In this study, a new automatic Arabic multi-document text summarization model is presented and discusses the structure of the proposed frameworks for multi-document Arabic text summarization. The most important steps in this research are document clustering and keyphrase extraction. We also describe important sentence extraction and filtering. Similarity algorithms, namely, cosine similarity and the Jaccard coefficient, are used to choose one sentence from each set of similar sentences and ignoring the rest (eliminate redundancy). The sentences are used to represent the text summary. The major contribution which can be advocated by the current study is a new model for automatic Arabic multi-document summarization, which is represented by using combined clustering method and keyphrase extraction. The combined clustering is a new approach of clustering which combines between hierarchical clustering methods and k-means clustering. The results of experiments proved the proposed model gives better performance in comparison to other works.

Table 1: Overall ROUGE-1 evaluation of the first, second and third experiment

Measure methods	CM	KC	KJ (%)	SL	CL
First experiment					
Recall	45.5	40.5	40.5	45.2	46.6
Precision	37.5	38.8	38.4	34.6	33.5
ROUGE-1	40.7	39.6	39.4	39.6	39.0
Second experiment					
Recall	45.2	32.3	42.3	37.0	43.4
Precision	41.8	56.6	39.6	48.6	40.0
ROUGE-1	43.5	41.0	40.9	42.0	41.7
Third experiment					
Recall	41.7	36.5	35.3	37.0	46.6
Precision	44.2	46.8	47.2	42.9	45.0
ROUGE-1	42.9	41.0	40.4	39.8	39.2

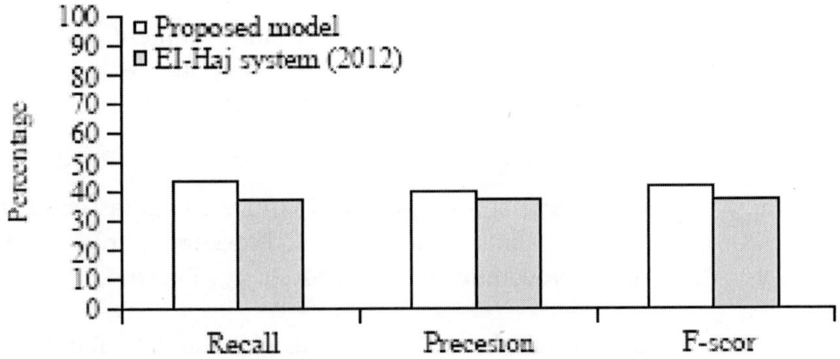

Figure 2: ROUGE-1: Comparison the proposed model with El-Haj (2012)

ACKNOWLEDGMENT

Our sincere thanks to Universiti Kebangsaan Malaysia (UKM), which have supported this study. Also, we would like to acknowledge and thank the General Directorate of Education in Qadisiyah, Ministry of Education, Republic of Iraq, for contributing in this study.

REFERENCES

1. Cimiano, P., A. Hotho and S. Staab, 2005. Learning concept hierarchies from text corpora using formal concept analysis. J. Artif. Intell. Res., 24: 305-339.

2. Conroy, J.M., J.D. Schlesinger, D.P. O'Leary and J. Goldstein, 2006. Back to basics: CLASSY 2006. Proceedings of the 6th Document Understanding Conferences, November 2006, New York -.

3. Douzidia, F.S. and G. Lapalme, 2004. Lakhas, an arabic summarising system. Proceedings of the 4th Document Understanding Conferences, May 2004, Rochester, pp: 128-135.

4. El-Haj, M., 2012. Multi-document arabic text summarisation. Ph.D. Thesis, University of Essex, UK.

5. El-Haj, M., U. Kruschwitz and C. Fox, 2011. Multi-document arabic text summarisation. Proceedings of the 3rd Computer Science and Electronic Engineering Conference, July 13-14, 2011, Colchester, UK., pp: 40-44.

6. El-Haj, M., U. Kruschwitz and C. Fox, 2011. University of essex at the TAC 2011 multilingual summarisation pilot. Proceedings of the Text Analysis Conference, November 14-15, 2011, Pilot, Maryland, USA -.

7. Fiszman, M., D. Demner-Fushman, H. Kilicoglu and T.C. Rindflesch, 2009. Automatic summarization of MEDLINE citations for evidence-based medical treatment: A topic-oriented evaluation. J. Biomed. Inform., 42: 801-813.

8. Giannakopoulos, G., V. Karkaletsis, G. Vouros and P. Stamatopoulos, 2008. Summarization system evaluation revisited: N-gram graphs. ACM Trans. Speech Language Process., Vol. 5. 10.1145/1410358.1410359

9. Hartigan, J.A., 1975. Clustering Algorithms. Books on Demand, New York, USA., ISBN-13: 9780608300498, Pages: 365.

10. Hirao, T., M. Okumura, N. Yasuda and H. Isozaki, 2007. Supervised automatic evaluation for summarization with voted regression model. Inform. Process. Manage., 43: 1521-1535.

11. Huang, A., 2008. Similarity measures for text document clustering. Proceedings of the New Zealand Computer Science Research Student Conference, April 14-17, 2008, Christchurch, New Zealand, pp: 49-56. Jain, A.K. and R.C. Dubes, 1988. Algorithms for Clustering Data. Prentice Hall Inc., Englewood Cliffs, USA., ISBN: 0-13-022278-X, Pages: 320.

12. Kanungo, T., D.M. Mount, N.S. Netanyahu, C.D. Piatko, R.S. Angela and Y. Wu, 2002. An efficient k-means clustering algorithm: Analysis and implementation. IEEE Trans. Pattern Anal. Mach. Intell., 24: 881-892.

13. Lin, C.Y., 2004. A package for automatic evaluation of summaries. Proceedings of the Workshop on Text Summarization Branches Out, July 25-26, 2004, Barcelona, Spain, pp: 25-26.

14. Liu, X. and W.B. Croft, 2004. Cluster-based retrieval using language models. Proceedings of the 27th ACM SIGIR Conference on Research and Development in Information Retrieval, July 25-29, 2004, Sheffield, UK., pp: 186-193.

15. Luhn, H.P., 1958. The automatic creation of literature abstracts. IBM J. Res. Dev., 2: 159-165.

16. Vora, P. and B. Oza, 2013. A survey on k-mean clustering and particle swarm optimization. Int. J. Sci. Mod. Eng., 1: 24-26.

17. Willett, P., 1982. A comparison of some hierarchical agglomerative clustering algorithms for structure-property correlation. Analytica Chim. Acta, 136: 29-37.

Chapter 3

The SP Theory of Intelligence: An Overview

J Gerard Wolff

CognitionResearch.org, Menai Bridge, UK

ABSTRACT

This article is an overview of the *SP theory of intelligence*, which aims to simplify and integrate concepts across artificial intelligence, mainstream computing and human perception and cognition, with information compression as a unifying theme. It is conceived of as a brain-like system that receives "New" information and stores some or all of it in compressed form as "Old" information; and it is realised in the form of a computer model, a first version of the *SP machine*. The matching and unification of patterns and the concept of *multiple alignment* are central ideas. Using heuristic techniques, the system builds multiple alignments that are "good" in terms of information compression. For each multiple alignment, probabilities may be calculated for associated inferences. Unsupervised learning is done by deriving new structures from partial matches between patterns and via heuristic search for sets of structures that are "good" in terms of information compression. These are normally ones that people judge to be "natural", in accordance with the "DONSVIC" principle—the discovery of natural structures via information compression. The SP theory provides an interpretation for concepts and phenomena in several other areas, including "computing", aspects of mathematics and logic, the representation of knowledge, natural language processing, pattern recognition, several kinds of reasoning, information storage and retrieval, planning and problem solving, information compression, neuroscience and human perception and cognition. Examples include the parsing and production of language with discontinuous dependencies in syntax, pattern recognition at multiple levels of abstraction and its integration with part-whole relations, nonmonotonic reasoning and reasoning with default values, reasoning in Bayesian networks, including "explaining away", causal diagnosis, and the solving of a geometric analogy problem.

KEYWORDS

information compression; artificial intelligence; multiple alignment; computing; representation of knowledge; natural language processing; pattern recognition; information retrieval; probabilistic reasoning; planning; problem solving; unsupervised learning

1. INTRODUCTION

The SP theory of intelligence, which has been under development since about 1987 [1], aims to simplify and integrate concepts across artificial intelligence, mainstream computing and human perception and cognition, with information compression as a unifying theme.

The name "SP" is short for *Simplicity* and *Power*, because compression of any given body of information, *I*, may be seen as a process of reducing informational "redundancy" in *I* and thus increasing its "simplicity", whilst retaining as much as possible of its non-redundant expressive "power". Likewise with Occam's Razor (Section 2.3, below).

Aspects of the theory, as it has developed, have been described in several peer-reviewed articles [2]. The most comprehensive description of the theory as it stands now, with many examples, is in [3]. But this book, with more than 450 pages, is too long to serve as an introduction to the theory. This article aims to meet that need, with a fairly full description of the theory and a selection of examples [4]. For the sake of brevity, the book will be referred to as "*BK*".

The next section describes the origins and motivation for the SP theory. Section 3 introduces the theory. Section 4 and Section 5 fill in a lot of the details, while the following sections describe aspects of the theory and what it can do.

2. ORIGINS AND MOTIVATION

The following subsections outline the origins of the SP theory, how it relates to some other research and how it has developed.

2.1. Information Compression

Much of the inspiration for the SP theory is a body of research, pioneered by Fred Attneave [5], Horace Barlow [6,7], and others, showing that several aspects of the workings of brains and nervous systems may be understood in terms of information compression [8].

For example, when we view a scene with two eyes, the image on the retina of the left eye is almost exactly the same as the image on the retina of right eye, but our brains merge the two images into a single percept and thus compress the information [7,9].

More immediately, the theory has grown out of my own research, developing models of the unsupervised learning of a first language, where the importance of information compression became increasingly clear (e.g., [10]. See also [11]).

The theory also draws on principles of "minimum length encoding" pioneered by Ray Solomonoff [12,13] and others. And it has become apparent that several aspects of computing, mathematics and logic may be understood in terms of information compression (*BK*, Chapters 2 and 10).

At an abstract level, information compression can bring two main benefits:

- For any given body of information, *I*, information compression may reduce its size and thus facilitate the storage, processing and transmission of *I*.

- Perhaps more important is the close connection between information compression and concepts of prediction and probability (see, for example, [14]). In the SP system, it is the basis for all kinds of inference and for calculations of probabilities.

In animals, we would expect these things to have been favoured by natural selection, because of the competitive advantage they can bring. Notwithstanding the "QWERTY" phenomenon—the dominance of the QWERTY keyboard despite its known inefficiencies—there is reason to believe that information compression, properly applied, may yield comparable advantages in artificial systems.

2.2. The Matching and Unification of Patterns

In the SP theory, the matching and unification of patterns is seen as being closer to the bedrock of information compression than more mathematical techniques such as wavelets or arithmetic coding, and closer to the bedrock of information processing and intelligence than, say, concepts of probability. A working hypothesis in this programme of research is that, by staying close to relatively simple, "primitive", concepts of matching patterns and unifying them, there is a better chance of cutting through unnecessary complexity and in gaining new insights and better solutions to problems. The mathematical basis of wavelets, arithmetic coding and probabilities may itself be founded on the matching and unification of patterns (*BK*, Chapter 10).

2.3. Simplification and Integration of Concepts

In accordance with Occam's Razor, the SP system aims to combine conceptual simplicity with descriptive and explanatory power. Apart from this widely-accepted principle, the drive for simplification and integration of concepts in this research programme has been motivated in part by Allen Newell's critique [15] of some kinds of research in cognitive science and, in part, by the apparent fragmentation of research in artificial intelligence and mainstream computing, with their myriad of concepts and many specialisms.

In attempting to simplify and integrate ideas, the SP theory belongs in the same tradition as unified theories of cognition, such as Soar [16] and ACT-R [17]—both of them inspired by Allen Newell [15]. Furthermore, it chimes with the resurgence of interest in understanding artificial intelligence as a whole (see, for example [18]) and with research on "natural computation" [19].

Although the SP programme shares some objectives with projects such as the Gödel Machine [20] and "universal artificial intelligence" [21], the approach is very different.

2.4. Transparency in the Representation of Knowledge

In this research, it is assumed that knowledge in the SP system should normally be transparent or comprehensible, much as in the "symbolic" tradition in artificial intelligence (see also, Section 5.2), and distinct from the kind of "sub-symbolic" representation of knowledge that is the rule in, for example, "neural networks" as they are generally conceived in computer science.

As we shall see in Section 7 and elsewhere in this article, SP patterns in the multiple alignment framework may serve to represent a variety of kinds of knowledge, in symbolic forms.

2.5. Development of the Theory

In developing the theory, it was apparent at an early stage that existing systems—such as my models of language learning [11] and systems like Prolog—would need radical re-thinking to meet the goal of simplifying and integrating ideas across a wide area.

The first published version of the SP theory [22] described "some unifying ideas in computing". Early work on the SP computer model concentrated on developing an improved version of the "dynamic programming" technique for the alignment of two sequences (see *BK*, Appendix A) as a possible route to modelling human-like flexibility in pattern recognition, analysis of language, and the like.

About 1992, it became apparent that the explanatory range of the theory could be greatly expanded by forming alignments of two, three or more sequences, much as in the "multiple alignment" concept of bioinformatics. That idea was developed and adapted in new versions of the SP model and incorporated in new procedures for unsupervised learning.

Aspects of the theory, with many examples, have been developed in the book [3].

3. INTRODUCTION TO THE SP THEORY

The main elements of the SP theory are:

- The SP theory is conceived as an abstract brain-like system that, in an "input" perspective, may receive "New" information via its senses and store some or all of it in its memory as "Old" information, as illustrated schematically in Figure 1. There is also an "output" perspective, described in Section 4.5.

- The theory is realised in the form of a computer model, introduced in Section 3.1, below, and described more fully later.

- All New and Old information is expressed as arrays (*patterns*) of atomic symbols in one or two dimensions. An example of an SP pattern may be seen in each row in Figure 4. Each symbol can be matched in an all-or-nothing manner with any other symbol. Any meaning that is associated with an atomic symbol or group of symbols must be expressed in the form of other atomic symbols.

- Each pattern has an associated frequency of occurrence, which may be assigned by the user or derived via the processes for unsupervised learning. The default value for the frequency of any pattern is 1.

- The system is designed for the unsupervised learning of Old patterns by compression of New patterns [23].

- An important part of this process is, where possible, the economical (compressed) encoding of New patterns in terms of Old patterns. This may be seen to achieve such things as pattern recognition, parsing or understanding of natural language, or other kinds of interpretation of incoming information in terms of stored knowledge, including several kinds of reasoning.

- In keeping with the remarks in Section 2.2, compression of information is achieved via the matching and unification (merging) of patterns. In this, there are key roles for the frequency of occurrence of patterns, and their sizes.

- The concept of *multiple alignment*, described in Section 4, is a powerful central idea, similar to the concept of multiple alignment in bioinformatics, but with important differences.

- Owing to the intimate connection, previously mentioned, between information compression and concepts of prediction and probability, it is relatively straightforward for the SP system to calculate probabilities for inferences made by the system and probabilities for parsings, recognition of patterns, and so on (Section 4.4).

- In developing the theory, I have tried to take advantage of what is known about the psychological and neurophysiological aspects of human perception and cognition and to ensure that the theory is compatible with such knowledge (see Section 14).

3.1. The SP Computer Model

The SP theory is realised most fully in the SP70 computer model, with capabilities in the building of multiple alignments and in unsupervised learning. This will be referred to as the SP model, although in some cases examples are from a subset of the model or slightly earlier precursors of it.

The SP model and its precursors have played a key part in the development of the theory:

- As an antidote to vagueness. As with all computer programs, processes must be defined with sufficient detail to ensure that the program actually works.

- By providing a convenient means of encoding the simple but important mathematics that underpins the SP theory, and performing relevant calculations, including calculations of probability.

- By providing a means of seeing quickly the strengths and weaknesses of proposed mechanisms or processes. Many ideas that looked promising have been dropped as a result of this kind of testing.
- By providing a means of demonstrating what can be achieved with the theory.

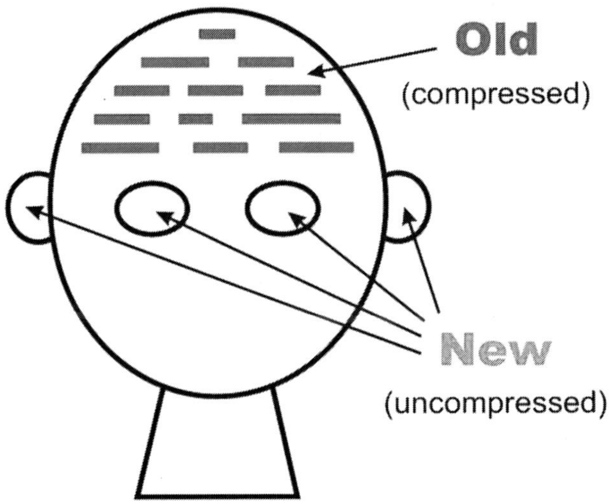

Figure 1. Schematic representation of the SP system from an "input" perspective.

The workings of the SP model are described in some detail in *BK* (Sections 3.9, 3.10, and 9.2) and more briefly in Section 4 andSection 5, below [24].

The two main elements of the models, described in Section 4 and Section 5, are the building of multiple alignments and the unsupervised learning of Old patterns.

3.2. The SP Machine

The SP computer model may be regarded as a first version of the *SP machine*, an expression of the SP theory and a means for it to be applied.

A useful step forward in the development of the SP theory would be the creation of a high-parallel, open-source version of the SP machine, accessible via the web, and with a good user interface [25]. This would provide a means for researchers to explore what can be done with the system and to improve it. How things may develop is shown schematically in Figure 2.

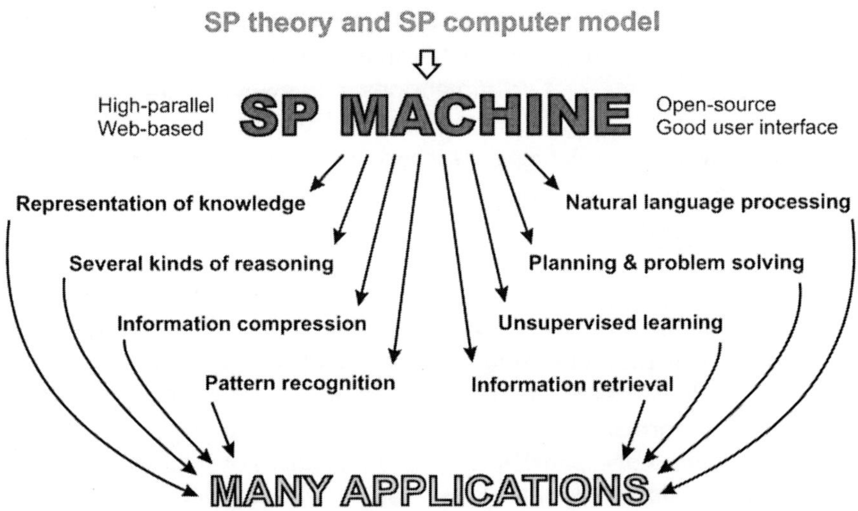

Figure 2. Schematic representation of the development and application of the proposed SP machine.

The high-parallel search mechanisms in any of the existing internet search engines would probably provide a good foundation for the proposed development.

Further ahead, there may be a case for the creation of new kinds of hardware, dedicated to the building of multiple alignments and other processes in the SP framework (Section 6.13 in [26]).

3.3. Unfinished Business

Like most theories, the SP theory has shortcomings, but it appears that they may be overcome. At present, the most immediate problems are:

- *Processing of information in two or more dimensions*. No attempt has yet been made to generalise the SP model to work with patterns in two dimensions, although that appears to be feasible to do, as outlined in *BK* (Section 13.2.1). As noted in *BK* (Section 13.2.2), it is possible that information with dimensions higher than two may be encoded in terms of patterns in one or two dimensions, somewhat in the manner of architects' drawings. A 3D structure may be stitched together from several partially-overlapping 2D views, in much the same way that, in digital photography, a panoramic view may be created from partially-overlapping pictures (Sections 6.1 and 6.2 in [27]).

- *Recognition of perceptual features in speech and visual images*. For the SP system to be effective in the processing of speech or visual images, it seems likely that some kind of preliminary processing will be required to identify low-level perceptual features, such as, in the case of speech, phonemes, formant ratios or formant transitions, or, in the case of visual images, edges, angles, colours, luminances or textures. In vision, at least, it seems likely that the SP framework itself will prove relevant, since edges may be seen as zones of non-redundant information between uniform areas containing more redundancy and, likewise, angles may be seen to provide significant information where straight edges, with more redundancy, come together (Section 3 in [27]). As a stop-gap solution, the preliminary processing may be done using existing techniques for the identification of low-level perceptual features (Chapter 13 in [28]).

- *Unsupervised learning*. A limitation of the SP computer model as it is now is that it cannot learn intermediate levels of abstraction in grammars (e.g., phrases and clauses), and it cannot learn the kinds of discontinuous dependencies in natural language syntax that are described in Section 8.1 to Section 8.3. I believe these problems are soluble and that solving them will greatly enhance the capabilities of the system for the unsupervised learning of structure in data (Section 5).

- *Processing of numbers*. The SP model works with atomic symbols, such as ASCII characters or strings of characters with no intrinsic meaning. In itself, the SP system does not recognise the arithmetic meaning of numbers such as “37” or “652” and will not process them correctly. However, the system has the potential to handle mathematical concepts if it is supplied with patterns representing Peano’s axioms or similar information (*BK*, Chapter 10). As a stop-gap solution in the SP machine, existing technologies may provide whatever arithmetic processing may be required.

4. THE MULTIPLE ALIGNMENT CONCEPT

The concept of *multiple alignment* in the SP theory has been adapted from a similar concept in bioinformatics, where it means a process of arranging, in rows or columns, two or more DNA sequences or amino-acid sequences, so that matching symbols—as many as possible—are aligned orthogonally in columns or rows.

Multiple alignments like these are normally used in the computational analysis of (symbolic representations of) sequences of DNA bases or sequences of amino acid residues as part of the process of elucidating the structure, functions or evolution of the corresponding molecules. An example of this kind of multiple alignment is shown in Figure 3.

```
  G G A     G     C A G G G A G G A     T G     G   G G A
  | | |     |     | | | | | | | | |     | |     |   | | |
  G G | G   G C C C A G G G A G G A     | G G C G   G G A
  | | |     | | | | | | | | | | | |     | |     |   | | |
A | G A C T G C C C A G G G | G G | G C T G     G A | G A
  | | |           | | | | | | | | |   |   |     |   | | |
  G G A A         | A G G G A G G A   | A G     G   G G A
  | |   |         | | | | | | | |     |   |     |   | | |
  G G C A         C A G G G A G G     C   G     G   G G A
```

Figure 3. A “good” alignment amongst five DNA sequences.

As in bioinformatics, a multiple alignment in the SP system is an arrangement of two or more patterns in rows (or columns), with one pattern in each row (or column) [29]. The main difference between the two concepts is that, in bioinformatics, all sequences have the same status, whereas in the SP theory, the system attempts to create a multiple alignment which enables one New pattern (sometimes more) to be encoded economically in terms of one or more Old patterns. Other differences are described in *BK* (Section 3.4.1).

In Figure 4, row 0 contains a New pattern representing a sentence: "t h i s b o y l o v e s t h a t g i r l", while each of rows 1 to 8 contains an Old pattern representing a grammatical rule or a word with grammatical markers. This multiple alignment, which achieves the effect of parsing the sentence in terms of grammatical structures, is the best of several built by the model when it is supplied with the New pattern and a set of Old patterns that includes those shown in the figure and several others as well.

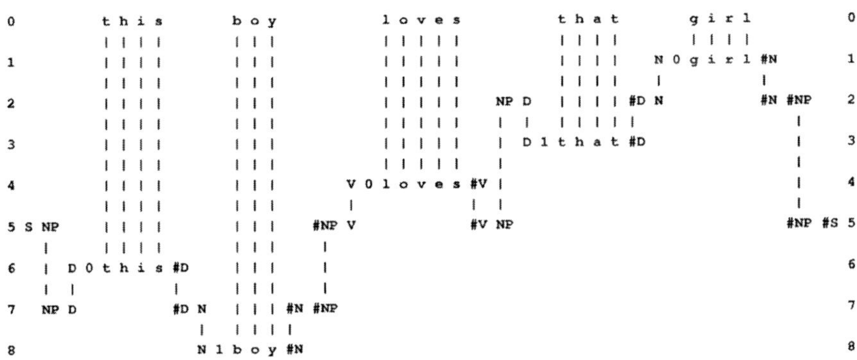

Figure 4. The best multiple alignment found by the SP model with the New pattern "t h i s b o y l o v e s t h a t g i r l" and a set of user-supplied Old patterns representing some of the grammatical forms of English, including words with their grammatical markers.

In this example, and others in this article, "best" means that the multiple alignment in the figure is the one that enables the New pattern to be encoded most economically in terms of the Old patterns, as described in Section 4.1, below.

4.1. Coding and the Evaluation of an Alignment in Terms of Compression

This section describes in outline how multiple alignments are evaluated in the SP model. More detail may be found in *BK* (Section 3.5).

Each Old pattern in the SP system contains one or more "identification" symbols, or *ID-symbols*, which, as their name suggests, serve to identify the pattern. Examples of ID-symbols in Figure 4 are "D" and "0" at the beginning of "D 0 t h i s #D" (row 6) and "N" and "1" at the beginning of "N 1 b o y #N" (row 8).

Associated with each type of symbol (where a "type" of symbol is any one of a set of symbols that match each other exactly) is a notional *code* or bit pattern that serves to distinguish the given type from all the others. This is only notional, because the bit patterns are not actually constructed. All that is needed for the purpose of evaluating multiple alignments is the size of the notional bit pattern associated with each type. This is calculated via the Shannon-Fano-Elias coding scheme (described in [30]), using information about the frequency of occurrence of each Old pattern, so that the shortest codes represent the most frequent symbol types and *vice versa*[31]. Notice that these bit patterns and their sizes are totally independent of the names for symbols that are used in written accounts like this one: names that are chosen purely for their mnemonic value.

Given a multiple alignment like the one shown in Figure 4, one can derive a *code pattern* from the multiple alignment in the following way:

1) Scan the multiple alignment from left to right looking for columns that contain an ID-symbol by itself, not aligned with any other symbol.

2) Copy these symbols into a code pattern in the same order that they appear in the multiple alignment.

The code pattern derived in this way from the multiple alignment shown in Figure 4 is "S 0 1 0 1 0 #S". This is, in effect, a compressed representation of those symbols in the New pattern that are aligned with Old symbols in the multiple alignment. In this case, the code pattern is a compressed representation of *all* the symbols in the New pattern, but it often happens that some of the symbols in the New pattern are not matched with any Old symbols and then the code pattern will represent only those New symbols that are aligned with Old symbols.

In the context of natural language processing, it is perhaps more plausible to suppose that the encoding of a sentence is some kind of representation of the meaning of the sentence, instead of a pattern like "S 0 1 0 1 0 #S". How a meaning may be derived from a sentence via multiple alignment is described in *BK* (Section 5.7).

4.1.1. Compression Difference and Compression Ratio

Given a code pattern, like "S 0 1 0 1 0 #S", we may calculate a "compression difference" as:

$$CD = B_N - B_E \tag{1}$$

or a "compression ratio" as:

$$CR = B_N / B_E \tag{2}$$

where B N is the total number of bits in those symbols in the New pattern that are aligned with Old symbols in the alignment and B Eis the total number of bits in the symbols in the code pattern, and the number of bits for each symbol is calculated via the Shannon-Fano-Elias scheme, as mentioned above.

C D and C R are each an indication of how effectively the New pattern (or those parts of the New pattern that are aligned with symbols within Old patterns in the alignment) may be compressed in terms of the Old patterns that appear in the given multiple alignment. The C D of a multiple alignment—which has been found to be more useful than C R —may be referred to as the*compression score* of the multiple alignment.

In each of these equations, B N is calculated as:

$$B_N = \sum_{i=1}^{h} C_i \tag{3}$$

where C i is the size of the code for the *i*th symbol in a sequence, H 1 . . . H h , comprising those symbols within the New pattern that are aligned with Old symbols within the multiple alignment.

B E is calculated as:

$$B_E = \sum_{i=1}^{s} C_i \quad (4)$$

where C i is the size of the code for the *i*th symbol in the sequence of *s* symbols in the code pattern derived from the multiple alignment.

4.2. The Building of Multiple Alignments

This section describes in outline how the SP model builds multiple alignments. More detail may be found in *BK* (Section 3.10).

Multiple alignments are built in stages, with pairwise matching and alignment of patterns. At each stage, any partially-constructed multiple alignment may be processed as if it was a basic pattern and carried forward to later stages. This is broadly similar to some programs for the creation of multiple alignments in bioinformatics [32]. At all stages, the aim is to encode New information economically in terms of Old information and to weed out multiple alignments that score poorly in that regard.

The model may create Old patterns for itself, as described in Section 5, but when the formation of multiple alignments is the focus of interest, Old patterns may be supplied by the user. In all cases, New patterns must be supplied by the user.

At each stage of building multiple alignments, the operations are as follows:

1) Identify a set of "driving" patterns and a set of "target" patterns. At the beginning, the New pattern is the sole driving pattern, and the Old patterns are the target patterns. In all subsequent stages, the best of the multiple alignments formed so far (in terms of their C D scores) are chosen to be driving patterns, and the target patterns are the Old patterns together with a selection of the best multiple alignments formed so far, including all of those that are driving patterns.

2) Compare each driving pattern with each of the target patterns to find full matches and good partial matches between patterns. This is done with a process that is essentially a form of "dynamic programming"

[33], somewhat like the WinMerge utility for finding similarities and differences between files [34]. The process is described quite fully in *BK* (Appendix A) and outlined in Section 4.2.1, below. The main difference between the SP process and others is that the former can deliver several alternative matches between a pair of patterns, while WinMerge and standard methods for finding alignments deliver one "best" result.

3) From the best of the matches found in the current stage, create corresponding multiple alignments and add them to the repository of multiple alignments created by the program.

This process of matching driving patterns against target patterns and building multiple alignments is repeated until no more multiple alignments can be found. For the best of the multiple alignments created since the start of processing, probabilities are calculated, as described in Section 4.4.

4.2.1. Finding Good Matches between Patterns

Figure 5 shows with a simple example how the SP model finds good full and partial matches between a "query" string of atomic symbols (alphabetic characters in this example) and a "database" string:

1) The query is processed left to right, one symbol at a time.

2) Each symbol in the query is, in effect, broadcast to every symbol in the database to make a yes/no match in each case.

3) Every positive match (hit) between a symbol from the query and a symbol in the database is recorded in a *hit structure*, illustrated in the figure.

4) If the memory space allocated to the hit structure is exhausted at any time, then the hit structure is purged: the leaf nodes of the tree are sorted in reverse order of their probability values, and each leaf node in the bottom half of the set is extracted from the hit structure, together with all nodes on its path which are not shared with any other path. After the hit structure has been purged, the recording of hits may continue using the space, which has been released.

4.2.2. Noisy Data

Because of the way each model searches for a global optimum in the building of multiple alignments, it does not depend on the presence or absence of any particular feature or combination of features. Up to a point, plausible results may be obtained in the face of errors of omission, commission and substitution in the data. This is illustrated in the two multiple alignments in Figure 6, where the New pattern in row 0 of (b) is the same sentence as in (a) ("t w o k i t t e n s p l a y"), but with the omission of the "w" in "t w o", the substitution of "m" for "n" in "k i t t e n s" and the addition of "x" within the word "p l a y". Despite these errors, the best multiple alignment created by the SP model is, as shown in (b), the one that we judge intuitively to be "correct".

This kind of ability to cope gracefully with noisy data is very much in keeping with our ability to understand speech in noisy surroundings, to understand written language despite errors and to recognise people, trees, houses and the like, despite fog, snow, falling leaves, or other things that may obstruct our view. In a similar way, it is likely to prove useful in artificial systems for such applications as the processing of natural language and the recognition of patterns.

4.3. Computational Complexity

In considering the matching and unification of patterns, it not hard to see that, for any body of information *I*, except very small examples, there is a huge number of alternative ways in which patterns may be matched against each other, there will normally be many alternative ways in which patterns may be unified, and exhaustive search is not tractable (*BK*, Section 2.2.8.4).

However, with the kinds of heuristic techniques that are familiar in other artificial-intelligence applications—reducing the size of the search space by pruning the search tree at appropriate points, and being content with approximate solutions which are not necessarily perfect—this kind of matching becomes quite practical. Much the same can be said about the heuristic techniques used for the building of multiple alignments (Section 4.2) and for unsupervised learning (Section 5.1).

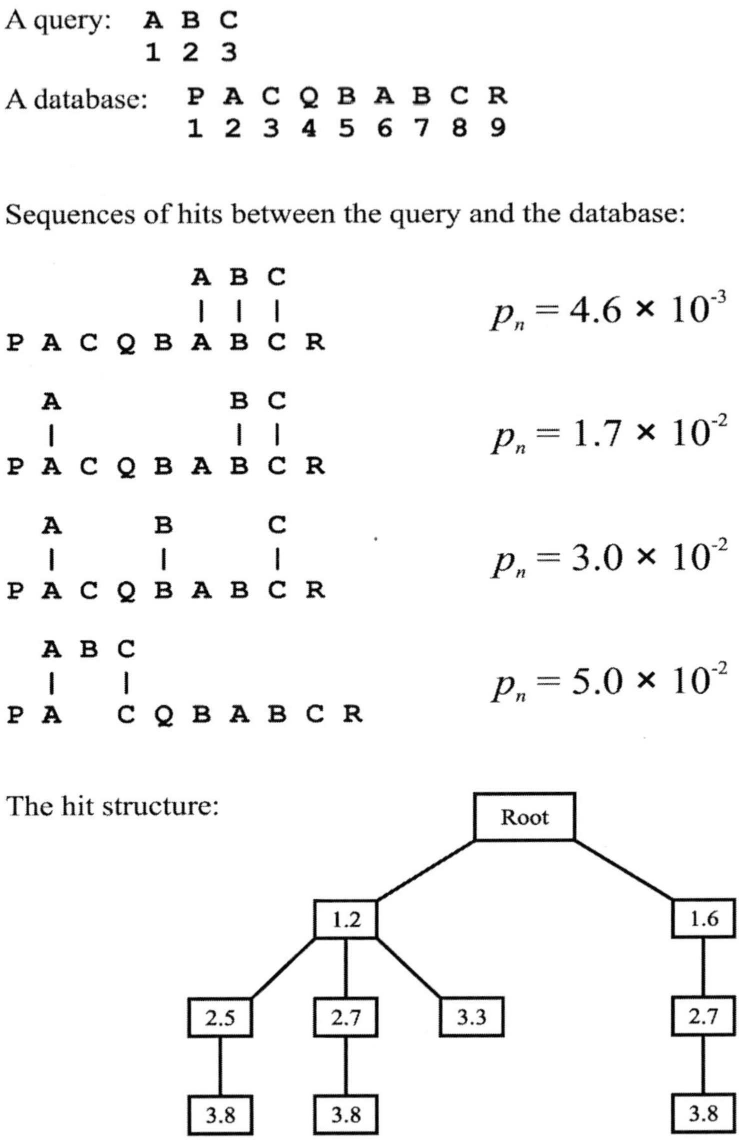

Figure 5. An example to show how the SP model finds good full and partial matches between patterns. A "query" string and a "database" string are shown at the top with the ordinal positions of symbols marked. Sequences of hits between the query and the database are shown in the middle with corresponding values of p n (described in *BK*, Section A.2). Each node in the hit structure shows the ordinal position of a query symbol and the ordinal position of a matching database symbol. Each path from the root node to a leaf node represents a sequence of hits.

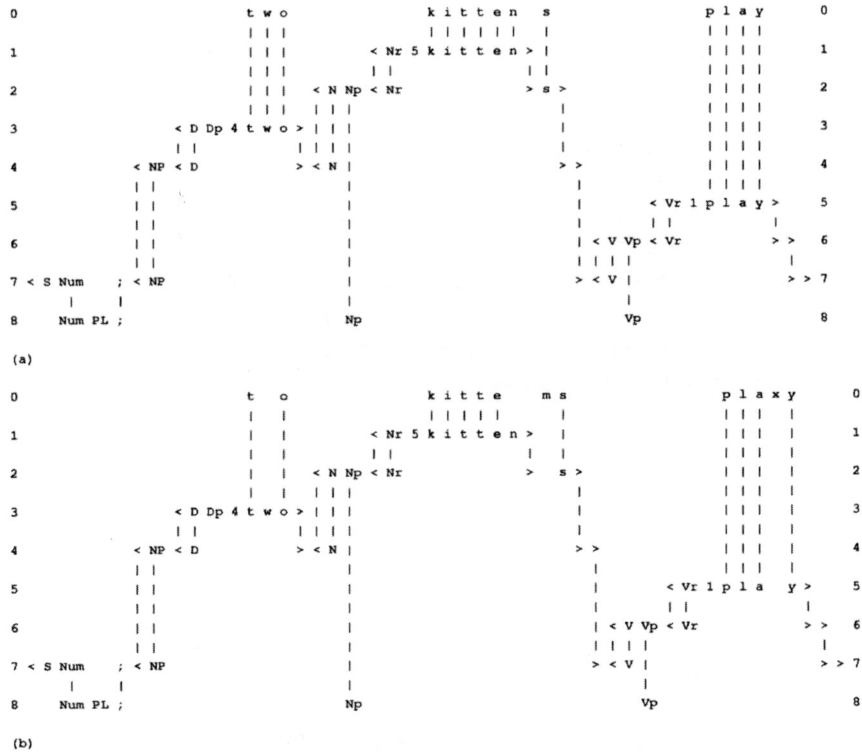

Figure 6. (**a**) The best multiple alignment created by the SP model with a store of Old patterns, like those in rows 1 to 8 (representing grammatical structures, including words) and a New pattern (representing a sentence to be parsed) shown in row 0; (**b**) as in (a), but with errors of omission, commission and substitution, and with the same set of Old patterns as before. (a) and (b) are reproduced from Figures 1 and 2 in [35], with permission.

An example of how effective this rough-and-ready approach can be is the way colonies of ants can find reasonably good solutions to the travelling salesman problem via the simple technique of marking their routes with pheromones and choosing routes that are most strongly marked [36].

For the process of building multiple alignments in the SP model, the time complexity in a serial processing environment, with conservative assumptions, has been estimated to be O (log 2 n × n m) , where *n* is the size of the pattern from New (in bits) and *m* is the sum of the lengths of the patterns in Old (in bits). In a parallel processing environment, the time complexity may approach O (log 2n × n) , depending on how well the parallel processing is applied. In serial and parallel environments, the space complexity has been estimated to be O (m) .

Although the data sets used with the current SP model have generally been small—because the main focus has been on the concepts being modelled and not the speed of processing—there is reason to be confident that the models can be scaled up to deal with large data sets because the kind of flexible matching of patterns, which is at the heart of the SP model, is done very fast and with huge volumes of data by all the leading internet search engines. As was suggested in Section 3.2, the relevant processes in any one of those search engines would probably provide a good basis for the creation of a high-parallel version of the SP machine.

4.4. Calculation of Probabilities Associated with Multiple Alignments

As described in *BK* (Chapter 7), the formation of multiple alignments in the SP framework supports several kinds of probabilistic reasoning. The core idea is that any Old symbol in a multiple alignment that is *not* aligned with a New symbol represents an inference that may be drawn from the multiple alignment. This section outlines how probabilities for such inferences may be calculated. There is more detail in *BK* (Section 3.7).

4.4.1. Absolute Probabilities

Any sequence of *L* symbols, drawn from an alphabet of | A | alphabetic types, represents one point in a set of *N* points, where *N* is calculated as:

$$N = |A|^L \tag{5}$$

If we assume that the sequence is random or nearly so (see *BK*, Section 3.7.1.1)—which means that the *N* points are equi-probable or nearly so—the probability of any one point (which represents a sequence of length *L*) is close to:

$$p_{ABS} = |A|^{-L} \quad (6)$$

This equation may be used to calculate the absolute probability of the code pattern that may be derived from any given multiple alignment (as described in Section 4.1). That number may also be regarded as the absolute probability of any inferences that may be drawn from the multiple alignment. In this calculation, *L* is the sum of all the bits in the symbols of the code pattern, and | A | is 2.

As we shall see (Section 4.4.3), Equation 6 may, with advantage, be generalised by replacing *L* with a value, L g e n , calculated in a slightly different way.

4.4.2. Relative Probabilities

The absolute probabilities of multiple alignments, calculated as described in the last subsection, are normally very small and not very interesting in themselves. From the standpoint of practical applications, we are normally interested in the *relative* values of probabilities, calculated as follows.

1) For the multiple alignment which has the highest C D (which we shall call the *reference multiple alignment*), identify the*reference set of symbols in New*, meaning the symbols from New which are encoded by the multiple alignment.

2) Compile a *reference set of multiple alignments*, which includes the reference multiple alignment and all other multiple alignments (if any), which encode exactly the reference set of symbols from New, neither more nor less.

3) Calculate the sum of the values for p A B S in the reference set of multiple alignments:

$$p_{A\_SUM} = \sum_{i=1}^{i=R} p_{ABS_i} \quad (7)$$

where *R* is the size of the reference set of multiple alignments and p A B S i is the value of p A B S for the *i*th multiple alignment in the reference set.

4) For each multiple alignment in the reference set, calculate its relative probability as:

$$p_{REL_i} = p_{ABS_i} / p_{A\_SUM} \qquad (8)$$

The values of p R E L , calculated as just described, provide an effective means of comparing the multiple alignments in the reference set.

4.4.3. A Generalisation of the Method for Calculating Absolute and Relative Probabilities

The value of *L*, calculated as described in Section 4.4.1, may be regarded as the informational "cost" of encoding the New symbol or symbols that appear in the multiple alignment, excluding those New symbols that have *not* appeared in the multiple alignment.

This is OK, but it is somewhat restrictive, because it means that if we want to calculate relative probabilities for two or more multiple alignments, they must all encode the same symbol or symbols from New. We cannot easily compare multiple alignments that encode different New symbols.

The generalisation proposed here is that, in the calculation of absolute probabilities, a new value, L g e n , would be used instead of*L*. This would be calculated as:

$$Lgen = L + Nnot \qquad (9)$$

where *L* is the total number of bits in the symbols in the code patterns (as in Section 4.4.1) and N n o t is the total number of bits in the New symbols that have *not* appeared in the multiple alignment.

The rationale is that, to encode *all* the symbols in New, we can use the code pattern to encode those New symbols that do appear in the multiple alignment, and for each of the remaining New symbols, we can simply use its code. The advantage of this scheme is that we can compare any two or more multiple alignments, regardless of the number of New symbols that appear in the multiple alignment.

4.4.4. Relative Probabilities of Patterns and Symbols

It often happens that a given pattern from Old, or a given symbol type within patterns from Old, appears in more than one of the multiple alignments in the reference set. In cases like these, one would expect the relative probability of the pattern or symbol type to be higher than if it appeared in only one multiple alignment. To take account of this kind of situation, the SP model calculates relative probabilities for individual patterns and symbol types in the following way:

1) Compile a set of patterns from Old, each of which appears at least once in the reference set of multiple alignments. No single pattern from Old should appear more than once in the set.

2) For each pattern, calculate a value for its relative probability as the sum of the p R E L values for the multiple alignments in which it appears. If a pattern appears more than once in a multiple alignment, it is only counted once for that multiple alignment.

3) Compile a set of symbol types, which appear anywhere in the patterns identified in step 2.

4) For each alphabetic symbol type identified in step 3, calculate its relative probability as the sum of the relative probabilities of the patterns in which it appears. If it appears more than once in a given pattern, it is only counted once.

The foregoing applies only to symbol types, which do not appear in the New. Any symbol type that appears in the New necessarily has a probability of one—because it has been observed, not inferred.

4.5. One System for Both the Analysis and the Production of Information

A potentially useful feature of the SP system is that the processes which serve to analyse or parse a New pattern in terms of Old patterns and to create an economical encoding of the New pattern, may also work in reverse, to recreate the New pattern from its encoding. This is the "output" perspective, mentioned in Section 3.

If the New pattern is the code sequence "S 0 1 0 1 0 #S" (as described in Section 4), and if the Old patterns are the same as were used to create the multiple alignment shown in Figure 4, then the best multiple alignment found by the system is the one shown inFigure 7. This multiple alignment contains the same words as the original sentence ("t h i s b o y l o v e s t h a t g i r l"), in the same order as the original. Readers who are familiar with Prolog will recognise that this process of recreating the original sentence from its encoding is similar in some respects to the way in which an appropriately-constructed Prolog program may be run "backwards", deriving "data" from "results".

How is it possible to decompress the compressed code for the original sentence by using information compression? This apparent paradox—decompression by compression—may be resolved by ensuring that, when a code pattern like "S 0 1 0 1 0 #S" is used to recreate the original data, each symbol is treated, at least notionally, as if it contained a few more bits of information than is strictly necessary. That residual redundancy allows the system to recreate the original sentence by the same process of compression as was used to create the original parsing and encoding [37].

This process of creating a relatively large pattern from a relatively small encoding provides a model for the creation of sentences by a person or an artificial system. However, instead of the New pattern being a rather dry code, like "S 0 1 0 1 0 #S", it would be more plausible if it were some kind of representation of the meaning of the sentence, like that mentioned in Section 4.1. How a sentence may be generated from a representation of meaning is outlined in *BK* (Section 5.7.1).

Similar principles may apply to other kinds of "output", such as planning an outing, cooking a meal, and so on.

5. UNSUPERVISED LEARNING

As was mentioned in Section 2.1, part of the inspiration for the SP theory has been a programme of research developing models of the unsupervised learning of language. But although the SNPR model [38] is quite successful in deriving plausible grammars from samples of English-like artificial language, it has proved to be quite unsuitable as a basis for the SP theory. In order to accommodate other aspects of intelligence, such as pattern recognition, reasoning, and problem solving, it has been necessary to develop an entirely New conceptual framework, with multiple alignment at centre stage.

```
0 S      0              1                0                   1              0                #S 0
  |      |              |                |                   |              |                |
1 S NP   |              |          #NP V |           #V NP   |              |            #NP #S 1
    |    |              |           |  | |           |  |    |              |             |
2   |    |              |           |  V 0 l o v e s #V |    |              |             |     2
    |    |              |           |                   |    |              |             |
3   |    |              |           |                   |    |            N 0 g i r l #N  |     3
    |    |              |           |                   |    |            |           |   |
4   |    |              |           |                   NP D |         #D N           #N #NP    4
    |    |              |           |                      | |         |
5   |    |              |           |                      D 1 t h a t #D                       5
    |    |              |           |
6   |  D 0 t h i s #D   |           |                                                           6
    |  |           |    |           |
7   NP D           #D N |       #N #NP                                                          7
                      | |       |
8                     N 1 b o y #N                                                              8
```

Figure 7. The best multiple alignment found by the SP model with the New pattern, "S 0 1 0 1 0 #S", and the same Old patterns as were used to create the multiple alignment shown in Figure 4.

Therefore, there is now the curious paradox that, while the SP theory is rooted in work on unsupervised learning and that kind of learning has a central role in the theory, the SP model does much the same things as the earlier model and with similar limitations (Section 3.3 and Section 5.1.4). However, I believe that the New conceptual framework has many advantages, that it provides a much sounder footing for further developments, and that with some reorganisation of the learning processes in the SP computer model, its current weaknesses may be overcome (Section 5.1.4).

5.1. Outline of Unsupervised Learning in the SP Model

The outline of the SP model in this section aims to provide sufficient detail for a good intuitive grasp of how it works. A lot more detail may be found in *BK* (Chapter 9).

In addition to the processes for building multiple alignments, the SP model has processes for deriving Old patterns from multiple alignments, evaluating sets of newly-created Old patterns in terms of their effectiveness for the economical encoding of the New information, and the weeding out low-scoring sets. The system does not merely record statistical information, it uses that information to learn new structures.

5.1.1. Deriving Old Patterns from Multiple Alignments

The process of deriving Old patterns from multiple alignments is illustrated schematically in Figure 8. As was mentioned in Section 3, the SP system is conceived as an abstract brain-like system that, in "input" mode, may receive "New" information via its senses and store some or all of it as "Old" information. Here, we may think of it as the brain of a baby who is listening to what people are saying. Let us imagine that he or she hears someone say "t h a t b o y r u n s" [39]. If the baby has never heard anything similar, then, if it is stored at all, that New information may be stored as a relatively straightforward copy, something like the Old pattern shown in row 1 of the multiple alignment in part (a) of the figure.

```
0       t h a t g i r l r u n s     0
        | | | |         | | | |
1 A 1 t h a t b o y     r u n s #A 1

(a)

B 2 t h a t #B
C 3 b o y #C
C 4 g i r l #C
D 5 r u n s #D
E 6 B #B C #C D #D #E

(b)
```

Figure 8. (**a**) A simple multiple alignment from which, in the SP model, Old patterns may be derived; (**b**) Old patterns derived from the multiple alignment shown in (**a**).

Now, let us imagine that the information has been stored and that, at some later stage, the baby hears someone say "t h a t g i r l r u n s". Then, from that New information and the previously-stored Old pattern, a multiple alignment may be created like the one shown in part (a) of Figure 8. And, by picking out coherent sequences that are either fully matched or not matched at all, four putative words may be extracted: "t h a t", "g i r l", "b o y", and "r u n s", as shown in the first four patterns in part (b) of the figure. In each newly-created Old pattern, there are additional symbols, such as "B", "2" and "#B" that are added by the system and which serve to identify the pattern, to mark its boundaries, and to mark its grammatical category or categories.

In addition to these four patterns, a fifth pattern is created, "E 6 B #B C #C D #D #E", as shown in the figure, that records the sequence, "t h a t ... r u n s", with the category "C #C" in the middle representing a choice between "b o y" and "g i r l". Part (b) in the figure is the beginnings of a grammar to describe that kind of phrase.

5.1.2. Evaluating and Selecting Sets of Newly-Created Old Patterns

The example just described shows how Old patterns may be derived from a multiple alignment, but it gives a highly misleading impression of how the SP model actually works. In practice, the program forms many multiple alignments that are much less tidy than the one shown, and it creates many Old patterns that are clearly "wrong". However, the program contains procedures for evaluating candidate sets of patterns ("grammars") and weeding out those that score badly in terms of their effectiveness for encoding the New information economically. Out of all the muddle, it can normally abstract one or two "best" grammars, and these are normally ones that appear intuitively to be "correct", or nearly so. In general, the program can abstract one or more plausible grammars from a sample of English-like artificial language, including words, grammatical categories of words, and sentence structure.

In accordance with the principles of minimum length encoding [12,13], the aim of these processes of sifting and sorting is to minimise (G + E) , where *G* is the size (in bits) of the grammar that is under development and *E* is the size (in bits) of the New patterns when they have been encoded in terms of the grammar.

For a given grammar comprising patterns, $p_1 \ldots p_g$, the value of *G* is calculated as:

$$G = \sum_{i=1}^{i=g} (\sum_{j=1}^{j=L_i} s_j) \tag{10}$$

where L_i is the number of symbols in the *i*th pattern and s_j is the encoding cost of the *j*th symbol in that pattern.

Given that each grammar is derived from a set $a_1 \ldots a_n$ of multiple alignments (one multiple alignment for each pattern from the New), the value of *E* for the grammar is calculated as:

$$E = \sum_{i=1}^{i=n} e_i \tag{11}$$

where e_i is the size, in bits, of the code string derived from the *i*th multiple alignment (Section 4.1).

For a given set of patterns from New, a tree of alternative grammars is created with branching occurring wherever there are two or more alternative multiple alignments for a given pattern from New. The tree is grown in stages and pruned periodically to keep it within reasonable bounds. At each stage, grammars with high values for $(G + E)$ (which will be referred to as *T*) are eliminated.

5.1.3. Plotting Values for *G*, *E* and *T*

Figure 9 shows cumulative values for *G*, *E* and *T* as the SP model searches for good grammars for a succession of eight New patterns, each of which represents a sentence. Each point on each of the lower three graphs represents the relevant value (on the scale at the left) from the best grammar found after a given pattern from New has been processed. The graph labelled "*O*" shows cumulative values on the scale at the left for the succession of New patterns. The graph labelled "T/O" shows the amount of compression achieved (on the scale to the right).

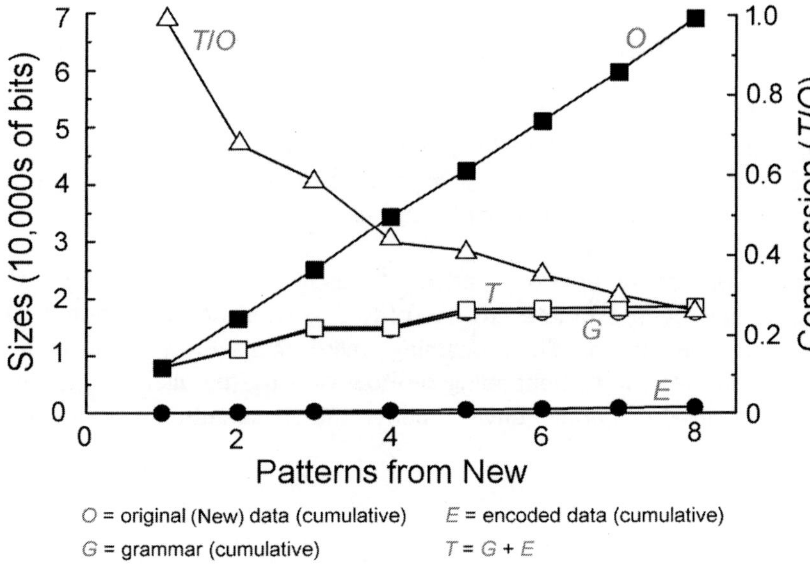

Figure 9. Changing values for *G*, *E* and *T* and related variables as learning proceeds, as described in the text.

5.1.4. Limitations in the SP Model and How They May Be Overcome

As mentioned before (Section 3.3), there are two main weaknesses in the processes for unsupervised learning in the SP model as it is now: the model does not learn intermediate levels in a grammar (phrases or clauses) or discontinuous dependencies of the kind described in Section 8.1 to Section 8.3.

It appears that some reorganisation of the learning processes in the model would solve both problems. What seems to be needed is a tighter focus on the principle that, with appropriately-constructed Old patterns, multiple alignments may be created without the kind of mismatch between patterns that may be seen in Figure 8 (a) ("g i r l" and "b o y" do not match each other), and that any such multiple alignment may be treated as if it was a simple pattern. That reform should facilitate the discovery of structures at multiple levels and the discovery of structures that are discontinuous in the sense that they can bridge intervening structures.

5.1.5. Computational Complexity

As with the building of multiple alignments (Section 4.3), the computational complexity of learning in the SP model is kept under control by pruning the search tree at appropriate points, aiming to discover grammars that are reasonably good and not necessarily perfect.

In a serial processing environment, the time complexity of learning in the SP model has been estimated to be O (N 2) , where *N* is the number of patterns in New. In a parallel processing environment, the time complexity may approach O (N) , depending on how well the parallel processing is applied. In serial or parallel environments, the space complexity has been estimated to be O (N) .

5.2. The Discovery of Natural Structures Via Information Compression (DONSVIC)

In our dealings with the world, certain kinds of structure appear to be more prominent and useful than others: in natural languages, there are words, phrase and sentences; we understand the visual and tactile worlds to be composed of discrete "objects"; and conceptually, we recognise classes of things, like "person", "house", "tree", and so on.

It appears that these "natural" kinds of structure are significant in our thinking because they provide a means of compressing sensory information, and that compression of information provides the key to their learning or discovery. At first sight, this looks like nonsense, because popular programs for compression of information, such as those based on the Lempel-Ziv-Welch (LZW) algorithm, or programs for JPEG compression of images, seem not to recognise anything resembling words, objects or classes. However, those programs are designed to work fast on low-powered computers. With other programs that are designed to be relatively thorough in their compression of information, natural structures can be revealed:

- Figure 10 shows part of a parsing of an unsegmented sample of natural language text created by the MK10 program [40] using only the information in the sample itself and without any prior dictionary or other knowledge about the structure of language. Although all spaces and punctuation had been removed from the sample, the program does reasonably well in revealing the word structure of the text. Statistical tests confirm that it performs much better than chance.

- The same program does quite well—significantly better than chance—in revealing phrase structures in natural language texts that have been prepared, as before, without spaces or punctuation—but with each word replaced by a symbol for its grammatical category [41]. Although that replacement was done by a person trained in linguistic analysis, the discovery of phrase structure in the sample is done by the program, without assistance.

- The SNPR program for grammar discovery [38] can, without supervision, derive a plausible grammar from an unsegmented sample of English-like artificial language, including the discovery of words, of grammatical categories of words, and the structure of sentences.

- In a similar way, with samples of English-like artificial languages, the SP model has demonstrated an ability to learn plausible structures, including words, grammatical categories of words and the structure of sentences.

It seems likely that the principles that have been outlined in this subsection may be applied not only to the discovery of words, phrases and grammars in language-like data, but also to such things as the discovery of objects in images [27] and classes of entity in all kinds of data. These principles may be characterised as *the discovery of natural structures via information compression*, or "DONSVIC" for short.

5.3. Generalisation, the Correction of Overgeneralisations and Learning from Noisy Data

Issues that arise in the learning of a first language, and probably in other kinds of learning, are illustrated in Figure 11:

- Given that we learn from a finite sample [42], represented by the smallest envelope in the figure, how do we generalise from that finite sample to a knowledge of the language corresponding to the middle-sized envelope, without overgeneralising into the region between the middle envelope and the outer one?

- How do we learn a "correct" version of our native language despite what is marked in the figure as "dirty data" (sentences that are not complete, false starts, words that are mispronounced, and more)?

Figure 10. Part of a parsing created by program MK10 [40] from a 10,000 letter sample of English (book 8A of the Ladybird Reading Series) with all spaces and punctuation removed. The program derived this parsing from the sample alone, without any prior dictionary or other knowledge of the structure of English. Reproduced from Figure 7.3 in [10], with permission.

One possible answer is that mistakes are corrected by parents, teachers and others. However, the weight of evidence is that children can learn their first language without that kind of assistance [43].

A better answer is the principle of minimum length encoding (described in its essentials in Section 5.1.2):

- As a general rule, the greatest reductions in (G + E) are achieved with grammars that represent moderate levels of generalisation, neither too little nor too much. In practice, the SNPR program, which is designed to minimise (G + E) , has been shown to produce plausible generalisations, without over-generalising [38].

- Any particular error is, by its nature, rare, and so, in the search for useful patterns (which, other things being equal, are the more frequently-occurring ones), it is discarded from the grammar along with other "bad" structures [44]. In the case of lossless compression, errors in any given body of data, *I*, would be retained in the encoding of *I*. However, with learning, it is normally the grammar and not the encoding that is the focus of interest. In practice, the MK10 and SNPR programs have been found to be quite insensitive to errors (of omission, addition or substitution) in their data, much as in the building of multiple alignments (Section 4.2.2).

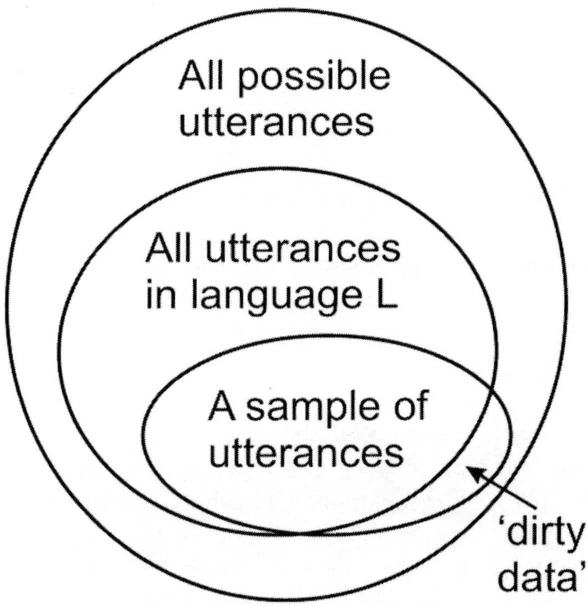

Figure 11. Categories of utterances involved in the learning of a first language, *L*. In ascending order size, they are: the finite sample of utterances from which a child learns; the (infinite) set of utterances in *L*; and the (infinite) set of all possible utterances. Adapted from Figure 7.1 in [10], with permission.

5.4. One-Trial Learning and Its Implications

In many theories of learning [45], the process is seen as gradual: behaviour is progressively shaped by rewards or punishments or other kinds of experience.

However, any theory of learning in which the process is necessarily gradual is out of step with our ordinary experience that we can and do learn things from a single experience, especially if that single experience is very significant for us (*BK*, Section 11.4.4.1).

In the SP theory, one-trial learning is accommodated in the way the system can store New information directly. And the gradual nature of, for example, language learning, may be explained by the complexity of the process of sifting and sorting the many alternative sets of candidate patterns to find one or more sets that are good in terms of information compression (*BK*, Section 11.4.4.2).

6. COMPUTING, MATHEMATICS AND LOGIC

Drawing mainly on *BK* (Chapters 4 to 11), this and the following sections describe, with a selection of examples, how the SP theory relates to several areas in artificial intelligence, mainstream computing, and human perception and cognition.

In *BK* (Chapter 4), I have argued that the SP system is equivalent to a universal Turing machine [46,47], in the sense that anything that may be computed with a Turing machine may, in principle, also be computed with an SP machine. The "in principle" qualification is necessary, because the SP theory is still not fully mature, and there are still some weaknesses in the SP computer model. The gist of the argument is that the operation of a post canonical system [48] may be understood in terms of the SP theory, and since it is accepted that the post canonical system is equivalent to the Turing machine (as a computational system), the Turing machine may also be understood in terms of the SP theory.

The key differences between the SP theory and earlier theories of computing are that the SP theory has a lot more to say about the nature of intelligence than earlier theories, that the theory is founded on principles of information compression via the matching and unification of patterns ("computing as compression"), and that it includes mechanisms for building multiple alignments and for heuristic search that are not present in earlier models.

6.1. Conventional Computing Systems

In conventional computing systems, compression of information may be seen in the matching of patterns with at least implicit unification of patterns that match each other—processes that appear in a variety of guises (*BK*, Chapter 2). And three basic techniques for the compression of information—*chunking-with-codes, schema-plus-correction* and *run-length coding*—may be seen in various forms in the organisation of computer programs (*ibid.*).

6.2. Mathematics and Logic

In a similar way, several structures and processes in mathematics and logic may be interpreted in terms of information compression via the matching and unification of patterns and the compression techniques just mentioned (*BK*, Chapter 10). For example: multiplication (as repeated addition) and exponentiation (as repeated multiplication) may be seen as examples of run-length coding; a function with parameters may be seen as an example of schema-plus-correction; the chunking-with-codes technique may be seen in the organisation of number systems; and so on.

6.3. Computing and Probabilities

As we have seen, the SP system is fundamentally probabilistic. If it is indeed Turing-equivalent, as suggested above, and if the Turing machine is regarded as a definition of "computing", then we may conclude that computing is fundamentally probabilistic. That may seem like a strange conclusion in view of the clockwork certainties that we associate with the operation of ordinary computers and the workings of mathematics and logic. There are at least three answers to that apparent contradiction:

- It appears that computing, mathematics and logic are more probabilistic than our ordinary experience of them might suggest. Gregory Chaitin has written: "I have recently been able to take a further step along the path laid out by Gödel and Turing. By translating a particular computer program into an algebraic equation of a type that was familiar even to the ancient Greeks, I have shown that there is randomness in the branch of pure mathematics known as number theory. My work indicates that—to borrow Einsteins metaphor—God sometimes plays dice with whole numbers." (p. 80 in [49]).

- The SP system may imitate the clockwork nature of ordinary computers by delivering probabilities of 0 and 1. This can happen with certain kinds of data, or tight constraints on the process of searching the abstract space of alternative matches, or both those things.

- It seems likely that the all-or-nothing character of conventional computers has its origins in the low computational power of early computers. In those days, it was necessary to apply tight constraints on the process of searching for matches between patterns. Otherwise, the computational demands would have been overwhelming. Similar things may be said about the origins of mathematics and logic, which have been developed for centuries without the benefit of any computational machine, except very simple and low-powered devices. Now that it is technically feasible to apply large amounts of computational power, constraints on searching may be relaxed.

7. REPRESENTATION OF KNOWLEDGE

Within the multiple alignment framework (Section 4), SP patterns may serve to represent several kinds of knowledge, including grammars for natural languages (Section 8; *BK*, Chapter 5), ontologies ([50]; *BK*, Section 13.4.3), class hierarchies with inheritance of attributes, including cross-classification or multiple inheritance (*BK*, Section 6.4), part-whole hierarchies and their integration with class-inclusion hierarchies (Section 9.1; *BK*, Section 6.4), decision networks and trees (*BK*, Section 7.5), relational tuples ([35]; *BK*, Section 13.4.6.1), if-then rules (*BK*, Section 7.6), associations of medical signs and symptoms [51], causal relations (*BK*, Section 7.9), and concepts in mathematics and logic, such as "function", "variable", "value", "set" and "type definition" (*BK*, Chapter 10).

The use of one simple format for the representation of knowledge facilitates the seamless integration of different kinds of knowledge.

8. NATURAL LANGUAGE PROCESSING

One of the main strengths of the SP system is in natural language processing (*BK*, Chapter 5):

- As illustrated in Figure 4, Figure 6 and Figure 7, grammatical rules, including words and their grammatical markers, may be represented with SP patterns.
- Both the parsing and production of natural language may be modelled via the building of multiple alignments (Section 4.5; *BK*, Section 5.7).
- The system can accommodate syntactic ambiguities in language (*BK*, Section 5.2) and also recursive structures (*BK*, Section 5.3).
- The framework provides a simple, but effective means of representing discontinuous dependencies in syntax (Section 8.1 toSection 8.3, below; *BK*, Sections 5.4 to 5.6).
- The system may also model non-syntactic "semantic" structures, such as class-inclusion hierarchies and part-whole hierarchies (Section 9.1).
- Because there is one simple format for different kinds of knowledge, the system facilitates the seamless integration of syntax with semantics (*BK*, Section 5.7).
- The system is robust in the face of errors of omission, commission or substitution in data (Section 4.2.2 and Section 5.3).
- The importance of context in the processing of language [52] is accommodated in the way the system searches for a global best match for patterns: any pattern or partial pattern may be a context for any other.

8.1. Discontinuous Dependencies in Syntax

The way in which the SP system can record discontinuous dependencies in syntax may be seen in both of the two parsings in Figure 6. The pattern in row 8 of each multiple alignment records the syntactic dependency between the plural noun phrase ("t w o k i t t e n s"), which is the subject of the sentence—marked with "Np"—and the plural verb phrase ("p l a y")—marked with "Vp"—which belongs with it.

This kind of dependency is discontinuous, because it can bridge arbitrarily large amounts of intervening structure, such as, for example, "from the West" in a sentence, like "Winds from the West are strong".

This method of marking discontinuous dependencies can accommodate overlapping dependencies, such as number dependencies and gender dependencies in languages like French (*BK*, Section 5.4). It also provides a means of encoding the interesting system of overlapping and interlocking dependencies in English auxiliary verbs, described by Noam Chomsky in *Syntactic Structures* [53].

In that book, the structure of English auxiliary verbs is part of Chomsky's evidence in support of Transformational Grammar. Despite the elegance and persuasiveness of his arguments, it turns out that the structure of English auxiliary verbs may be described with non-transformational rules in, for example, Definite Clause Grammars [54], and also in the SP system, as outlined in the subsections that follow.

8.2. Two Quasi-Independent Patterns of Constraint in English Auxiliary Verbs

In English, the syntax for main verbs and the auxiliary verbs which may accompany them follows two quasi-independent patterns of constraint which interact in an interesting way.

The *primary constraints* may be expressed with this sequence of symbols:

```
M H B B V,
```

which should be interpreted in the following way:

- Each letter represents a category for a single word:
  - "M" stands for "modal" verbs, like "will", "can", "would", *etc.*
  - "H" stands for one of the various forms of the verb, "to have".
  - Each of the two instances of "B" stands for one of the various forms of the verb, "to be".
  - "V" stands for the main verb, which can be any verb, except a modal verb (unless the modal verb is used by itself).
- The words occur in the order shown, but any of the words may be omitted.
- Questions of "standard" form follow exactly the same pattern as statements, except that the first verb, whatever it happens to be ("M", "H", the first "B", the second "B" or "V"), precedes the subject noun phrase instead of following it.

Here are two examples of the primary pattern with all of the words included:

```
It will have been being washed
     M    H    B     B      V

Will it have been being washed?
 M        H    B     B      V
```

The *secondary constraints* are these:

- Apart from the modals, which always have the same form, the first verb in the sequence, whatever it happens to be ("H", the first "B", the second "B" or "V"), always has a "finite" form (the form it would take if it were used by itself with the subject).
- If an "M" auxiliary verb is chosen, then whatever follows it ("H", first "B", second "B" or "V") must have an "infinitive" form (*i.e.*, the "standard" form of the verb as it occurs in the context, "to ...", but without the word "to").

- If an "H" auxiliary verb is chosen, then whatever follows it (the first "B", the second "B" or "V") must have a past tense form, such as "been", "seen", "gone", "slept", "wanted", *etc*. In Chomsky's *Syntactic Structures* [53], these forms were characterised as *en* forms, and the same convention has been adopted here.

- If the first of the two "B" auxiliary verbs is chosen, then whatever follows it (the second "B" or "V") must have an *ing* form, e.g., "singing", "eating", "having", "being", *etc*.

- If the second of the two "B" auxiliary verbs is chosen, then whatever follows it (only the main verb is possible now) must have a past tense form (marked with *en*, as above).

- The constraints apply to questions in exactly the same way as they do to statements.

Figure 12 shows a selection of examples with the dependencies marked.

8.3. Multiple Alignments and English Auxiliary Verbs

Without reproducing all the detail in *BK* (Section 5.5), we can see from Figure 13 and Figure 14 how the primary and secondary constraints may be applied in the multiple alignment framework.

In each figure, the sentence to be analysed is shown as a New pattern in column 0. The primary constraints are applied via the matching of symbols in Old patterns in the remaining columns, with a consequent interlocking of the patterns, so that they recognise sentences of the form, "M H B B V", with options as described above.

In Figure 13 [55], the secondary constraints apply as follows:

- The first verb, "is", is marked as having the finite form (with the symbol "FIN" in columns 5 and 7). The same word is also marked as being a form of the verb "to be" (with the symbol "B" in columns 4, 5 and 6). Because of its position in the parsing, we know that it is an instance of the second "B" in the sequence "M H B B V".

```
           H------en   B2---------en
          ----    --   --         --
It  will  have  been  being  washed
    ----  ----  --      ---  ----
     M----inf   B1------ing   V
```

```
          B1------ing
          --      ---
Will  he  be  talking?
----      --  ----
 M-------inf   V
```

```
              V
            ------
They  have  finished
      ----        --
       H----------en
      fin
```

```
Are  they  gone?
---        ----
B2----------en
fin         V
```

```
         B1--------ing
         --        ---
Has  he  been  working?
---        --  ----
 H---------en   V
fin
```

Figure 12. A selection of example sentences in English with markings of dependencies between the verbs. *Key:* "M" = modal, "H" = forms of the verb "have", "B1" = first instance of a form of the verb "be", "B2" = second instance of a form of the verb "be", "V" = main verb, "fin" = a finite form, "inf" = an infinitive form, "en" = a past tense form, "ing" = a verb ending in "ing".

- The second verb, “washed”, is marked as being in the *en* category (with the symbol “EN” in columns 1 and 4).

- That a verb corresponding to the second instance of “B” must be followed by an *en* kind of verb is expressed by the pattern, “B XV EN”, in column 4.

```
0      1      2     3    4    5      6      7      8     9     10

                                                         S
                                                         ST
                                                   NP -- NP
                                                   SNG ------- SNG
it ----------------------------------------------- it
                                                   #NP - #NP
                                            X1 --------- X1
                                            2
                                     XB1 -- XB1
                              V ---- V
                         B -- B ---- B
                              SNG ---------------------------- SNG
                              FIN --------- FIN
                              0
is -------------------------- is
                              #V --- #V
                                     #XB1 - #XB1
                                            #X1 -------- #X1
                                            XR --------- #XR
                                            XB
                    XV - XV --------------- XV
       V ---------- V
       EN -------------- EN
       5
wash - wash
       ED --- ED
ed ---------- ed
       #ED -- #ED
       #V --------- #V
                    #S -------------------- #S --------- #S

0      1      2     3    4    5      6      7      8     9     10
```

Figure 13. The best alignment found by the SP model with “it is wash ed” in New (column 0) and a user-supplied grammar in Old.

```
0      1      2      3      4     5    6      7      8     9     10    11     12    13     14    15   16

                                                                 S
                                                                 Q
                                                     X1 -------- X1
                                                     0
                                              V ---- V
                                              M ---- M ----------------------------------- M
                                              0
will ---------------------------------------- will
                                              #V --- #V
                                                     #X1 ------- #X1
                                                           NP -- NP
                                                           SNG
it ------------------------------------------------------- it
                                                           #NP - #NP
                                                     XR -------- XR
                                                     XH --------------------- XH
                                                                       V ---- V
                                                                       H ---- H --------------- H
                                                                       INF --------------- INF
have ----------------------------------------------------------------- have
                                                                       #V --- #V
                                                                              #XH
                                                     XB --------------------- XB
                                       XB ---------- XB
                     XB1 ------------- XB1
              V ---- V
              B ---- B ----------------------------------------------------------------------------- B
              EN ------------------------------------------------------------------------------ EN
be ---------- be
       EN1 -- EN1
en --- en
       #EN1 - #EN1
              #V --- #V
                     #XB1 ------------ #XB1
                                  XV - XV ---------- XV -------------------------------------------- XV
                            V --- V
                            EN --------------------------------------------------------------------- EN
                            1
brok ---------------------- brok
                            EN1 --------------------------------------------------- EN1
en -------------------------------------------------------------------------------- en
                            #EN1 ---------------------------------------------------#EN1
                            #V --- #V
                                  #S - #S ---------- #S -------- #S --------- #S

0      1      2      3      4     5    6      7      8     9     10    11     12    13     14    15   16
```

Figure 14. The best alignment found by the SP model with "will it have be en brok en" in New (column 0) and the same grammar in Old as was used for the example in Figure 13.

In Figure 14, the secondary constraints apply like this:

- The first verb, "will", is marked as modal (with "M" in columns 7, 8 and 14).
- The second verb, "have", is marked as having the infinitive form (with "INF" in columns 11 and 14), and it is also marked as a form of the verb, "to have" (with "H" in columns 11, 12, and 15).

- That a modal verb must be followed by a verb of infinitive form is marked with the pattern, "M INF", in column 14.

- The third verb, "been", is marked as being a form of the verb, "to be" (with "B" in columns 2, 3 and 16). Because of its position in the parsing, we know that it is an instance of the second "B" in the sequence, "M H B B V". This verb is also marked as belonging to the *en* category (with "EN" in columns 2 and 15).

- That an "H" verb must be followed by an "EN" verb is marked with the pattern, "H EN", in column 15.

- The fourth verb, "broken", is marked as being in the *en* category (with "EN" in columns 4 and 16).

- That a "B" verb (second instance) must be followed by an "EN" verb is marked with the pattern, "B XV EN", in column 16.

9. PATTERN RECOGNITION

The system also has some useful features as a framework for pattern recognition (*BK*, Chapter 6):

- It can model pattern recognition at multiple levels of abstraction, as described in *BK* (Section 6.4.1), and with the integration of class-inclusion relations with part-whole hierarchies (Section 9.1; *BK*, Section 6.4.1).

- The SP system can accommodate "family resemblance" or *polythetic* categories, meaning that recognition does not depend on the presence absence of any particular feature or combination of features. This is because there can be alternatives at any or all locations in a pattern and, also, because of the way the system can tolerate errors in data (next point).

- The system is robust in the face of errors of omission, commission or substitution in data (Section 4.2.2).

- The system facilitates the seamless integration of pattern recognition with other aspects of intelligence: reasoning, learning, problem solving, and so on.

- A probability may be calculated for any given identification, classification or associated inference (Section 4.4).

- As in the processing of natural language (Section 8), the importance of context in recognition [56] is accommodated in the way the system searches for a global best match for patterns. As before, any pattern or partial pattern may be a context for any other.

One area of application is medical diagnosis, viewed as pattern recognition [51]. There is also potential to assist in the understanding of natural vision and in the development of computer vision, as discussed in [27].

9.1. Class Hierarchies, Part-Whole Hierarchies and Their Integration

A strength of the multiple alignment concept is that it provides a simple but effective vehicle for the representation and processing of class-inclusion hierarchies, part-whole hierarchies and their integration.

Figure 15 shows the best multiple alignment found by the SP model with the New pattern, "white-bib eats furry purrs" (column 0), representing some features of an unknown creature, and with a set of Old patterns representing different classes of animal, at varying levels of abstraction. From this multiple alignment, we may conclude that the unknown entity is an animal (column 1), a mammal (column 2), a cat (column 3), and the specific individual "Tibs" (column 4).

The framework also provides for the representation of heterarchies or cross classification: a given entity, such as "Jane" (or a class of entities), may belong in two or more higher-level classes that are not themselves hierarchically related, such as "woman" and "doctor" [57].

The way that class-inclusion relations and part-whole relations may be combined in one multiple alignment is illustrated in Figure 16. Here, some features of an unknown plant are expressed as a set of New patterns, shown in column 0: the plant has chlorophyll, the stem is hairy, it has yellow petals, and so on.

```
0           1            2              3                  4

                                                           T
                                                           Tibs
                                        C ---------------- C
                                        cat
                         M ------------ M
                         mammal
            A ---------- A
            animal
            head ---------------------- head
                                        carnassial-teeth
            #head --------------------- #head
            body ----------------------------------------- body
white-bib ------------------------------------------------ white-bib
            #body ---------------------------------------- #body
            legs ---------------------- legs
                                        retractile-claws
            #legs --------------------- #legs
eats ------ eats
            breathes
            has-senses
            ...
            #A --------- #A
furry ------------------ furry
                         warm-blooded
                         ...
                         #M ----------- #M
purrs --------------------------------- purrs
                                        ...
                                        #C --------------- #C
                                                           tabby
                                                           ...
                                                           #T

0           1            2              3                  4
```

Figure 15. The best multiple alignment found by the SP model, with the New pattern "white-bib eats furry purrs" and a set of Old patterns representing different categories of animal and their attributes.

```
0                 1                2                  3              4             5                  6

                  <species>
                  acris
                  <genus> --------------------------------------------------------------------------- <genus>
                  Ranunculus ------------------------------------------------------------------------ Ranunculus
                                                                                   <family> --------- <family>
                                                                                   Ranunculaceae ---- Ranunculaceae
                                                                     <order> ------ <order>
                                                                     Ranunculales - Ranunculales
                                                      <class> ------ <class>
                                                      Angiospermae - Angiospermae
                                   <phylum> --------- <phylum>
                                   Plants ----------- Plants
                                   <feeding>
has_chlorophyll ------------------ has_chlorophyll
                                   photosynthesises
                                   <feeding>
                                   <structure> ------ <structure>
                                                      <shoot>
<stem> ---------- <stem> ---------------------------- <stem>
hairy ----------- hairy
</stem> --------- </stem> --------------------------- </stem>
                  <leaves> -------------------------- <leaves>
                  compound
                  palmately_cut
                  </leaves> ------------------------- </leaves>
                                                      <flowers> ------------------ <flowers>
                                                                                   <arrangement>
                                                                                   regular
                                                                                   all_parts_free
                                                                                   </arrangement>
                  <sepals> ------------------------------------------------------- <sepals>
                  not_reflexed
                  </sepals> ------------------------------------------------------ </sepals>
<petals> -------- <petals> ------------------------------------------------------- <petals> --------- <petals>
                                                                                   <number> --------- <number>
                                                                                                      five
                                                                                   </number> -------- </number>
                  <colour> ------------------------------------------------------- <colour>
yellow ---------- yellow
                  </colour> ------------------------------------------------------ </colour>
</petals> ------- </petals> ------------------------------------------------------ </petals> -------- </petals>
                                                                                   <hermaphrodite>
<stamens> ------------------------------------------------------------------------ <stamens>
numerous ------------------------------------------------------------------------- numerous
</stamens> ----------------------------------------------------------------------- </stamens>
                                                                                   <pistil>
                                                                                   ovary
                                                                                   style
                                                                                   stigma
                                                                                   </pistil>
                                                                                   </hermaphrodite>
                                                      </flowers> ----------------- </flowers>
                                                      </shoot>
                                                      <root>
                                                      </root>
                                   </structure> ----- </structure>
<habitat> ------- <habitat> ------ <habitat>
meadows --------- meadows
</habitat> ------ </habitat> ----- </habitat>
                  <common_name> -- <common_name>
                  Meadow
                  Buttercup
                  </common_name> - </common_name>
                                   <food_value> ---------------------------------- <food_value>
                                                                                   poisonous
                                   </food_value> --------------------------------- </food_value>
                                   </phylum> -------- </phylum>
                                                      </class> ----- </class>
                                                                     </order> ----- </order>
                                                                                   </family> -------- </family>
                  </genus> -------------------------------------------------------------------------- </genus>
                  </species>

0                 1                2                  3              4             5                  6
```

Figure 16. The best multiple alignment created by the SP model, with a set of New patterns (in column 0) that describe some features of an unknown plant and a set of Old patterns, including those shown in columns 1 to 6, that describe different categories of plant, with their parts and sub-parts and other attributes.

From this multiple alignment, we can see that the unknown plant is most likely to be the Meadow Buttercup, *Ranunculus acris*, as shown in column 1. As such, it belongs in the genus *Ranunculus* (column 6), the family Ranunculaceae (column 5), the order Ranunculales (column 4), the class Angiospermae (column 3), and the phylum Plants (column 2).

Each of these higher-level classifications contributes information about attributes of the plant and its division into parts and sub-parts. For example, as a member of the class Angiospermae (column 3), the plant has a shoot and roots, with the shoot divided into stem, leaves and flowers; as a member of the family Ranunculaceae (column 5), the plant has flowers that are "regular", with all parts "free"; as a member of the phylum Plants (column 2), the buttercup has chlorophyll and creates its own food by photosynthesis; and so on.

9.2. Inference and Inheritance

In the example just described, we can infer from the multiple alignment, very directly, that the plant, which has been provisionally identified as the Meadow Buttercup, performs photosynthesis (column 2), has five petals (column 6), is poisonous (column 5), and so on. As in other object-oriented systems, the first of these attributes has been "inherited" from the class "Plants", the second from the class *Ranunculus* and the third from the class Ranunculaceae. These kinds of inference illustrate the close connection, often noted, between pattern recognition and inferential reasoning (see also [58]).

10. PROBABILISTIC REASONING

The SP system can model several kinds of reasoning including inheritance of attributes (as just described), one-step "deductive" reasoning, abductive reasoning, reasoning with probabilistic decision networks and decision trees, reasoning with "rules", nonmonotonic reasoning and reasoning with default values, reasoning in Bayesian networks (including "explaining away"), causal diagnosis, and reasoning that is not supported by evidence (*BK*, Chapter 7).

Since these several kinds of reasoning all flow from one computational framework (multiple alignment), they may be seen as aspects of one process, working individually or together without awkward boundaries.

Plausible lines of reasoning may be achieved, even when relevant information is incomplete.

Probabilities of inferences may be calculated, including extreme values (0 or 1) in the case of logic-like "deductions".

A selection of examples is described in the following subsections.

10.1. Nonmonotonic Reasoning and Reasoning with Default Values

Conventional deductive reasoning is *monotonic*, because deductions made on the strength of current knowledge cannot be invalidated by new knowledge: the conclusion that "Socrates is mortal", deduced from "All humans are mortal" and "Socrates is human", remains true for all time, regardless of anything we learn later. By contrast, the inference that "Tweety can probably fly" from the propositions that "Most birds fly" and "Tweety is a bird" is *nonmonotonic* because it may be changed if, for example, we learn that Tweety is a penguin.

This section presents some examples that show how the SP system can accommodate nonmonotonic reasoning.

10.1.1. Typically, Birds Fly

The idea that (all) birds can fly may be expressed with an SP pattern, like "Bd bird name #name canfly warm-blooded wings feathers ... #Bd". This, of course, is an oversimplification of the real-world facts, because, while it true that the majority of birds fly, we know that there are also flightless birds, like ostriches, penguins and kiwis.

In order to model these facts more closely, we need to modify the pattern that describes birds to be something like this: "Bd bird name #name f #f warm-blooded wings feathers ... #Bd. And, to our database of Old patterns, we need to add patterns like this:

```
Default Bd f canfly #f #Bd #Default
P penguin Bd f cannotfly #f #Bd ... #P
O ostrich Bd f cannotfly #f #Bd ... #O.
```

Now, the pair of symbols, "f #f" in "Bd bird name #name f #f warm-blooded wings feathers ... #Bd", functions like a "variable" that may take the value, "canfly", if a given class of birds can fly and "cannotfly" when a type of bird cannot fly. The pattern "P penguin Bd f cannotfly #f #Bd ... #P" shows that penguins cannot fly and, likewise, the pattern "O ostrich Bd f cannotfly #f #Bd ... #O" shows that ostriches cannot fly. The pattern "Default Bd f canfly #f #Bd #Default", which has a substantially higher frequency than the other two patterns, represents the default value for the variable which is "canfly". Notice that all three of these patterns contains the pair of symbols "Bd ... #Bd" showing that the corresponding classes are all subclasses of birds.

10.1.2. Tweety is a Bird, So, Probably, Tweety Can Fly

When the SP model is run with "bird Tweety" in New and the same patterns in Old as before, modified as just described, the three best multiple alignments found are those shown in Figure 17 to Figure 19.

The first multiple alignment tells us that, with a relative probability of 0.66, Tweety may be the typical kind of bird that can fly. The second multiple alignment tells us that, with a relative probability of 0.22, Tweety might be an ostrich and, as such, he or she would not be able to fly. Likewise, the third multiple alignment tells us that, with a relative probability of 0.12, Tweety might be a penguin and would not be able to fly. The values for probabilities in this simple example are derived from guestimated frequencies that are, almost certainly, not ornithologically correct.

10.1.3. Tweety Is a Penguin, So Tweety Cannot Fly

Figure 20 shows the best multiple alignment found by the SP model when it is run again, with "penguin Tweety" in New instead of "bird Tweety". This time, there is only one multiple alignment in the reference set, and its relative probability is 1.0. Correspondingly, all inferences that we can draw from this multiple alignment have a probability of 1.0. In particular, we can be confident, within the limits of the available knowledge, that Tweety cannot fly.

In a similar way, if Tweety were an ostrich, we would be able to say with confidence ($p = 1.0$) that he or she would not be able to fly.

```
0          1          2                3

                                       Default
                      Bd ------------- Bd
bird ---------------- bird
           name ----- name
Tweety --- Tweety
           #name ---- #name
                      f -------------- f
                                       canfly
                      #f ------------- #f
                      warm-blooded
                      wings
                      feathers
                      ...
                      #Bd ------------ #Bd
                                       #Default

0          1          2                3
```

Figure 17. The first of the three best multiple alignments formed by the SP model with “bird Tweety” in New and patterns in Old, as described in the text. The relative probability of this multiple alignment is 0.66.

10.2. Reasoning in Bayesian Networks, Including "Explaining Away"

A Bayesian network is a directed, acyclic graph, like the one shown in Figure 21, below, where each node has zero or more "inputs" (connections with nodes that can influence the given node) and one or more "outputs" (connections to other nodes that the given node can influence).

Each node contains a set of conditional probability values, each one the probability of a given output value for a given input value or combination of input values. With this information, conditional probabilities of alternative outputs for any node may be computed for any given *combination* of inputs. By combining these calculations for sequences of nodes, probabilities may be propagated through the network from one or more "start" nodes to one or more "finishing" nodes.

This section describes how the SP system may perform that kind of probabilistic reasoning and some advantages compared with Bayesian networks.

Judea Pearl (p. 7 in [59]) describes the phenomenon of "explaining away" like this: "If A implies B, C implies B, and B is true, then finding that C is true makes A *less* credible. In other words, finding a second explanation for an item of data makes the first explanation less credible." (his italics). Here is an example:

Normally an alarm sound alerts us to the possibility of a burglary. If somebody calls you at the office and tells you that your alarm went off, you will surely rush home in a hurry, even though there could be other causes for the alarm sound. If you hear a radio announcement that there was an earthquake nearby and if the last false alarm you recall was triggered by an earthquake, then your certainty of a burglary will diminish.

The causal relationships in the example just described may be captured in a Bayesian network, like the one shown in Figure 21.

Pearl argues that, with appropriate values for conditional probabilities, the phenomenon of "explaining away" can be explained in terms of this network (representing the case where there is a radio announcement of an earthquake) compared with the same network without the node for "radio announcement" (representing the situation where there is no radio announcement of an earthquake).

```
0                1       2                 3

                                           O
                                           ostrich
                         Bd -------------- Bd
bird ------------------- bird
                 name -- name
Tweety - Tweety
                 #name - #name
                         f --------------- f
                                           cannotfly
                         #f -------------- #f
                         warm-blooded
                         wings
                         feathers
                         ...
                         #Bd ------------- #Bd
                                           ...
                                           #O

0                1       2                 3
```

Figure 18. The second of the three best multiple alignments formed by the SP model with "bird Tweety" in New and patterns in Old, as described in the text. The relative probability of this multiple alignment is 0.22.

10.2.1. Representing Contingencies with Patterns and Frequencies

To see how this phenomenon may be understood in terms of the SP theory, consider, first, the set of patterns shown in Figure 22, which are to be stored in Old. The patterns in the figure show events which occur together in some notional sample of the "world" together with their frequencies of occurrence in the sample.

```
0              1              2                        3

                                                       P
                                                       penguin
                              Bd ----------- Bd
bird ------------ bird
               name --- name
Tweety - Tweety
               #name -- #name
                              f ------------ f
                                                       cannotfly
                              #f ----------- #f
                              warm-blooded
                              wings
                              feathers
                              ...
                              #Bd ---------- #Bd
                                                       ...
                                                       #P

0              1              2                        3
```

Figure 19. The last of the three best multiple alignments formed by the SP model with "bird Tweety" in New and patterns in Old, as described in the text. The relative probability of this multiple alignment is 0.12.

As with other knowledge-based systems, we shall assume that the "closed-world" assumption applies, so that the absence of any pattern may be taken to mean that the corresponding combination of events did not occur in the period when observations were made [60].

```
burglary alarm (1000)
earthquake alarm (20)
alarm phone_alarm_call (980)
earthquake radio_earthquake_announcement (40)
e1 earthquake e2 (40)
```

```
0       1       2            3

                             P
penguin --------------------- penguin
                Bd --------- Bd
                bird
        name --- name
Tweety -- Tweety
        #name -- #name
                f ---------- f
                             cannotfly
                #f --------- #f
                warm-blooded
                wings
                feathers
                ...
                #Bd -------- #Bd
                             ...
                             #P

0       1       2            3
```

Figure 20. The best multiple alignment formed by the SP model with "penguin Tweety" in New and patterns in Old, as described in the text. The relative probability of this multiple alignment is 1.0.

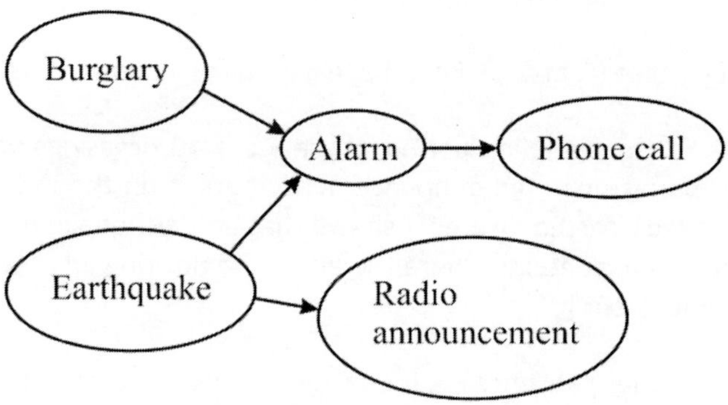

Figure 21. A Bayesian network representing causal relationships discussed in the text. In this diagram, "Phone call" means "a phone call about the alarm going off" and "Radio announcement" means "a radio announcement about an earthquake".

Figure 22. A set of patterns to be stored in Old in an example of "explaining away". The number in brackets after each pattern is the notional frequency of occurrence of the pattern. The symbol, "phone_alarm_call", is intended to represent a phone call conveying news that the alarm sounded; "radio_earthquake_announcement" represents an announcement on the radio that there has been an earthquake. The symbols "e1" and "e2" represent other contexts for "earthquake" besides the contexts "alarm" and "radio_earthquake_announcement".

The first pattern ("burglary alarm") shows that there were 1,000 occasions when there was a burglary and the alarm went off, and the second pattern ("earthquake alarm") shows just 20 occasions when there was an earthquake and the alarm went off (presumably triggered by the earthquake). Thus, we have assumed that, as triggers for the alarm, burglaries are much more common than earthquakes.

Since there is no pattern showing that the alarm sounded when there was a burglary and an earthquake at the same time, we may assume, via the closed-world assumption, that nothing like that happened during the sampling period.

The third pattern ("alarm phone_alarm_call") shows that, out of the 1,020 cases when the alarm went off, there were 980 cases where a phone call about the alarm was made. Since there is no pattern showing phone calls about the alarm in any other context, the closed-world assumption allows us to assume that there were no false positives (e.g., hoaxes)—phone calls about the alarm when no alarm had sounded.

The fourth pattern ("earthquake radio_earthquake_announcement")

shows that, in the sampling period, there were 40 occasions when there was an earthquake with an announcement about it on the radio. And the fifth pattern ("e1 earthquake e2") shows that an earthquake has occurred on 40 occasions in contexts where the alarm did not ring and there was no radio announcement [61].

As before, the absence of patterns, like "earthquake alarm radio_earthquake_announcement", representing cases where an earthquake triggers the alarm and also leads to a radio announcement, allows us to assume via the closed-world assumption that cases of that kind have not occurred in the sampling period.

10.2.2. Approximating the Temporal Order of Events

In these patterns and in the multiple alignments shown below, the left-to-right order of symbols may be regarded as an approximation to the order of events in time. Thus, in the first two patterns, events that can trigger an alarm precede the sounding of the alarm. Likewise, in the third pattern, "alarm" (meaning that the alarm has sounded) precedes "phone_alarm_call" (a phone call to say the alarm has sounded). A single dimension can only approximate the order of events in time, because it cannot represent events that overlap in time or which occur simultaneously. However, this kind of approximation has little or no bearing on the points to be illustrated here.

10.2.3. Other Considerations

Other points relating to the patterns shown in Figure 22 include:

- No attempt has been made to represent the idea that "the last false alarm you recall was triggered by an earthquake" (p. 9 in [59]). At some stage in the development of the SP system, there will be a need to take account of recency (*BK*, Section 13.2.6).
- With these imaginary frequency values, it has been assumed that burglaries (with a total frequency of occurrence of 1,160) are much more common than earthquakes (with a total frequency of 100). As we shall see, this difference reinforces the belief that there has been a burglary when it is known that the alarm has gone off (but without additional knowledge of an earthquake).
- In accordance with Pearl's example (p. 49 in [59]) (but contrary to the phenomenon of looting during earthquakes), it has been assumed that earthquakes and burglaries are independent. If there was some association between them, then, in accordance with the closed-world assumption, there should be a pattern in Figure 22 representing the association.

10.2.4. Formation of Alignments: The Burglar Alarm has Sounded

Receiving a phone call to say that the burglar alarm at one's house has gone off may be represented by placing the symbol, "phone_alarm_call", in New. Figure 23 shows, at the top, the best multiple alignment formed by the SP model in this case, with the patterns from Figure 22 in Old. The other two multiple alignments in the reference set are shown below the best multiple alignment, in order of C D value and relative probability. The actual values for C D and relative probability are given in the caption to Figure 23.

The unmatched Old symbols in these multiple alignments represent inferences made by the system. The probabilities for these inferences, which are calculated by the SP model (as outlined in Section 4.4) are shown in Table 1. These probabilities do not add up to 1 and we should not expect them to, because any given multiple alignment may contain two or more of these symbols.

The most probable inference is the rather trivial inference that the alarm has indeed sounded. This reflects the fact that there is no pattern in Figure 22 representing false positives for telephone calls about the alarm. Apart from the inference that the alarm has sounded, the most probable inference ($p = 0.328$) is that there has been a burglary. However, there is a distinct possibility that there has been an earthquake—but the probability in this case ($p = 0.016$) is much lower than the probability of a burglary.

These inferences and their relative probabilities seem to accord quite well with what one would naturally think following a telephone call to say that the burglar alarm at one's house has gone off (given that one was living in a part of the world where earthquakes were not vanishingly rare).

10.3. Formation of Alignments: The Burglar Alarm Has Sounded and There is a Radio Announcement of an Earthquake

In this example, the phenomenon of "explaining away" occurs when you learn not only that the burglar alarm has sounded, but that there has been an announcement on the radio that there has been an earthquake. In terms of the SP model, the two events (the phone call about the alarm and the announcement about the earthquake) can be represented in New by a pattern like this:

phone_alarm_call radio_earthquake_announcement

or "radio_earthquake_announcement phone_alarm_call". The order of the two symbols makes no difference to the result.

```
0       phone_alarm_call 0
               |
1 alarm phone_alarm_call 1

(a)

0                phone_alarm_call 0
                        |
1          alarm phone_alarm_call 1
             |
2 burglary alarm                  2

(b)

0                  phone_alarm_call 0
                          |
1            alarm phone_alarm_call 1
               |
2 earthquake alarm                  2

(c)
```

Figure 23. The best multiple alignment (at the top) and the other two multiple alignments in its reference set formed by the SP model with the pattern, "phone_alarm_call", in New and the patterns from Figure 22 in Old. In order from the top, the values for C D with relative probabilities in brackets are: 19.91 (0.656), 18.91 (0.328) and 14.52 (0.016).

Table 1. The probabilities of unmatched Old symbols, calculated by the SP model for the three multiple alignments shown in Figure 23.

Symbol	*Probability*
alarm	1.0
burglary	0.328
earthquake	0.016

In this case, there is only one multiple alignment (shown at the top of Figure 24) that can "explain" all the information in New. Since there is only this one multiple alignment in the reference set for the best multiple alignment, the associated probabilities of the inferences that can be read from the multiple alignment ("alarm" and "earthquake") are one: it was an earthquake that caused the alarm to go off (and led to the phone call) and not a burglary.

These results show how "explaining away" may be explained in terms of the SP theory. The main point is that the multiple alignment or multiple alignments that provide the best "explanation" of a telephone call to say that one's burglar alarm has sounded is different from the multiple alignment or multiple alignments that best explain the same telephone call coupled with an announcement on the radio that there has been an earthquake. In the latter case, the best explanation is that the earthquake triggered the alarm. Other possible explanations have lower probabilities.

10.3.1. Other Possibilities

As mentioned above, the closed-world assumption allows us to rule out possibilities, such as:

- A burglary (which triggered the alarm) and, at the same time, an earthquake (which led to a radio announcement) or

- An earthquake that triggered the alarm and led to a radio announcement and, at the same time, a burglary that did not trigger the alarm.

- Many other unlikely possibilities of a similar kind ([59], also discussed in Section 2.2.4 of this article).

Nevertheless, we may consider possibilities of that kind by combining multiple alignments as described in *BK* (Section 7.8.7). However, as a general rule, that kind of further analysis makes no difference to the original conclusion: *the multiple alignment, which was originally judged to represent the best interpretation of the available facts, has not been dislodged from this position*. This is in keeping with the way we normally concentrate on the most likely explanations of events and ignore the many conceivable, but unlikely, alternatives.

10.4. The SP framework and Bayesian Networks

The foregoing examples show that the SP framework is a viable alternative to Bayesian networks, at least in the kinds of situation that have been described. This subsection makes some general observations about the relative merits of the two frameworks for probabilistic reasoning where the events of interest are subject to multiple influences or chains of influence or both those things.

Without in any way diminishing Thomas Bayes' achievement, his theorem appears to have shortcomings as the basis for theorising about perception and cognition:

- *Undue complexity in the storage of statistical knowledge.* Each node in a Bayesian network contains a table of conditional probabilities for all possible combinations of inputs, and these tables can be quite large. By contrast, the SP framework only requires a single measure of frequency for each pattern. A focus on frequencies seems to yield an overall advantage in terms of simplicity compared with the representation of statistical knowledge in the form of conditional probabilities.

- *Diverting attention from simpler alternatives.* By emphasising probabilities, Bayes' theorem diverts attention away from simpler and more primitive concepts of matching and unification of patterns, which, by hypothesis, provide the foundation for several aspects of intelligence (Section 2.2).

- *No place for structural learning.* Bayes' theorem assumes that the objects and categories that are to be related to each other via conditional probabilities are already "given". It has nothing to say about how ontological knowledge may be created from raw perceptual input. By contrast, the SP framework provides for the discovery of objects and other categories via the matching and unification of patterns, in accordance with the DONSVIC principle (Section 5.2).

10.5. Causal Diagnosis

In this section, we consider a simple example of a fault diagnosis in an electronic circuit—described by [59] (pp. 263–272). Figure 25 shows the circuit with inputs on the left, outputs on the right and, in between, three multipliers (M 1 , M 2 and M 3) and two adders (M 4 and M 5). For the given inputs on the left, it is clear that output F is false and output G is correct.

Figure 26 shows a causal network derived from the electronic circuit in Figure 25 (p. 264 in [59]). In this diagram, *X, Y, Z, F* and *G*represent the outputs of components M 1 , M 2 , M 3 , M 4 and M 5 , respectively. In each case, there are three causal influences on the output: the two inputs to the component and the state of the component, which may be "good" or "bad". These influences are shown in Figure 26 by lines with arrows connecting the source of the influence to the target node. Thus, for example, the two inputs of component M 1 are represented by *A* and *C*, the good or bad state of component M 1 is represented by the node labelled M 1 and their causal influences on node *X* are shown by the three arrows pointing at that node.

Given a causal analysis like this, and given appropriate information about conditional probabilities, it is possible to derive one or more alternative diagnoses of which components are good and which are bad. In Pearl's example, it is assumed that components of the same type have the same prior probability of failure and that the probability of failure of multipliers is greater than for adders. Given these assumptions and some others, together with the inputs and outputs shown in Figure 25 (but not the intermediate values), the best diagnosis derived from the causal network is that the M 1 component is bad and the second best diagnosis is that M 4 is bad. Pearl indicates that some third-best interpretations may be retrievable (e.g., M 2 and M 5 are bad) "... but in general, it is not guaranteed that interpretations beyond the second-best will be retrievable." (p. 272 in [59]).

```
0                       phone_alarm_call radio_earthquake_announcement 0
                                |                    |
1                 alarm phone_alarm_call             |                 1
                    |                                |
2 earthquake alarm                                   |                 2
       |                                             |
3 earthquake                         radio_earthquake_announcement 3

(a)

0 phone_alarm_call radio_earthquake_announcement 0
                                |
1 earthquake       radio_earthquake_announcement 1

(b)

0        phone_alarm_call radio_earthquake_announcement 0
               |
1 alarm phone_alarm_call                                1

(c)

0                     phone_alarm_call radio_earthquake_announcement 0
                             |
1               alarm phone_alarm_call                               1
                  |
2 burglary alarm                                                     2

(d)

0                       phone_alarm_call radio_earthquake_announcement 0
                               |
1                 alarm phone_alarm_call                               1
                    |
2 earthquake alarm                                                     2

(e)
```

Figure 24. At the top, the best multiple alignment formed by the SP model with the pattern, "phone_alarm_call radio_earthquake_announcement", in New and the patterns from Figure 22 in Old. Other multiple alignments formed by the SP model are shown below. From the top, the C D values are: 74.64, 54.72, 19.92, 18.92 and 14.52.

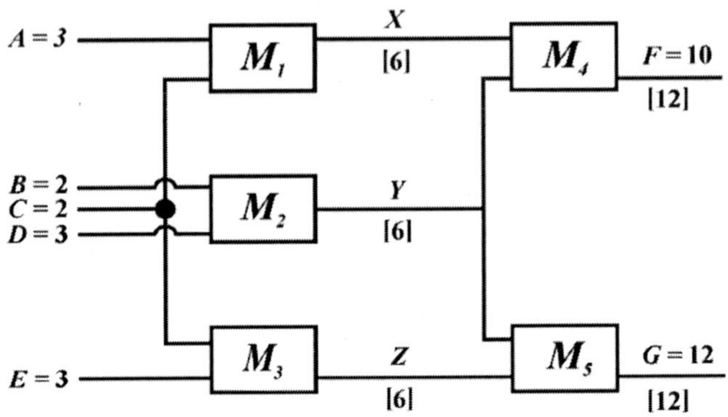

Figure 25. An electronic circuit containing three multipliers, M 1 , M 2 and M 3 , and two adders, M 4 and M 5 . Correct outputs are shown in square brackets. Redrawn from Figure 5.6 in [59](p. 263), with permission from the copyright holder, Elsevier.

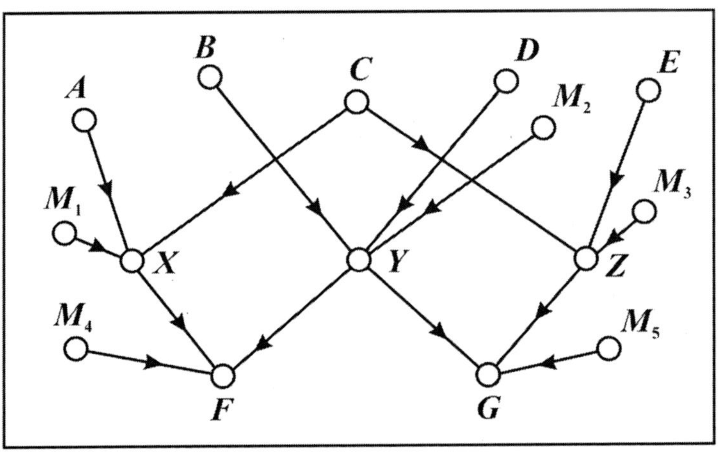

Figure 26. A causal network derived from the electronic circuit in Figure 25. Redrawn from Figure 5.7 in [59](p. 264), with permission from the copyright holder, Elsevier.

10.6. An SP Approach to Causal Diagnosis

The main elements of the SP analysis presented here are as follows:

- The input-output relations of any component may be represented as a set of patterns, each one with a measured or estimated frequency of occurrence.

- With suitable extensions, these patterns may serve to transfer the output of one component to the input of another.

- A "framework" pattern (shown at the bottom of Figure 27) is needed to ensure that appropriate multiple alignments can be built.

Figure 27 shows a set of patterns for the circuit shown in Figure 25. In the figure, the patterns that start with the symbol "M1" represent input-output relations for component M 1 , those that start with "M2" represent input-output relations for the M 2 component and, likewise, for the other patterns except the last one (starting with the symbol "frame"), which is the framework pattern mentioned above. For each initial symbol, there is a corresponding "terminating" symbol with an initial "#" character. For reasons explained shortly, there may be other symbols following the "terminating" symbol.

Let us now consider the first pattern in the figure ("M1 M1GOOD TM1I1 TM1I2 TM1O #M1 TM4I2") representing input-output relations for component M 1 when that component is good, as indicated by the symbol "M1GOOD". In this pattern, the symbols, "TM1I1", "TM1I2" and "TM1O", represent the two inputs and the output of the component, "#M1" is the terminating symbol and "TM4I2" serves to transfer the output of M 1 to the second input of component M 4 , as will be explained. If a symbol, like "TM1I1", "T", indicates that the input is true, "M1" identifies the component and "I1" indicates that this is the first input of the component. Other symbols may be interpreted in a similar way, following the key given in the caption of Figure 27. In effect, this pattern says that, when the component is working correctly, true inputs yield a true output. The pattern has a relatively high frequency of occurrence (500,000), reflecting the idea that the component will normally work correctly.

The other two patterns for component M 1 ("M1 M1BAD TM1I1 TM1I2 TM1O #M1 TM4I2" and "M1 M1BAD TM1I1 TM1I2 FM1O #M1 FM4I2") describe input-output relations when the component is bad. The first one describes the situation where true inputs to a faulty component yield a true result, a possibility noted by Pearl (*ibid.* p. 265). The second pattern—with a higher frequency—describes the more usual situation, where true inputs to a faulty component yield a false result. Both these bad patterns have much lower frequencies than the good pattern.

The other patterns in Figure 27 may be interpreted in a similar way. Components M 1 , M 2 and M 3 have only three patterns each, because it is assumed that inputs to the circuit will always be true, so it is not necessary to include patterns describing what happens when one or both of the inputs are false. By contrast, there are four good patterns and eight bad patterns for each of M 4 and M 5 , because either of these components may receive faulty input.

For each of the five components, the frequencies of the bad patterns sum to 100. However, for each of the components, M 1 , M 2and M 3 , the total frequency of the good patterns is 500,000 compared with 1,000,000 for the set of good patterns associated with each of the components, M 4 and M 5. These figures accord with the assumptions in Pearl's example that components of the same type have the same probability of failure and that the probability of failure of multipliers (M 1 , M 2 , and M 3) is greater than the probability of the failure of adders (M 4 and M 5).

10.7. Multiple Alignments in Causal Diagnosis

Given appropriate patterns, the SP model constructs multiple alignments from which diagnoses may be obtained. Figure 28 shows the best multiple alignment created by the SP model, with the Old patterns shown in Figure 27, and "TM1I1 TM1I2 TM2I1 TM2I2 TM3I1 TM3I2 FM4O TM5O" as New pattern. The first six symbols in this pattern express the idea that all the inputs for componentsM 1 , M 2 and M 3 are true. The penultimate symbol ("FM4O") shows that the output of M 4 is false, and the last symbol ("TM5O") shows that the output of M 5 is true—in accordance with the outputs shown in Figure 25.

From the multiple alignment in Figure 28, it can be inferred that component M 1 is bad and all the other components are good. A total of seven alternative diagnoses can be derived from those multiple alignments created by the SP model that encode all the symbols in New. These diagnoses are shown in Table 2, each with its relative probability.

```
M1 M1GOOD TM1I1 TM1I2 TM1O #M1 TM4I2 (500000)
M1 M1BAD TM1I1 TM1I2 TM1O #M1 TM4I2 (4)
M1 M1BAD TM1I1 TM1I2 FM1O #M1 FM4I2 (96)
M2 M2GOOD TM2I1 TM2I2 TM2O #M2 TM4I1 TM5I2 (500000)
M2 M2BAD TM2I1 TM2I2 TM2O #M2 TM4I1 TM5I2 (4)
M2 M2BAD TM2I1 TM2I2 FM2O #M2 FM4I1 FM5I2 (96)
M3 M3GOOD TM3I1 TM3I2 TM3O #M3 TM5I1 (500000)
M3 M3BAD TM3I1 TM3I2 TM3O #M3 TM5I1 (4)
M3 M3BAD TM3I1 TM3I2 FM3O #M3 FM5I1 (96)
M4 M4GOOD TM4I1 TM4I2 TM4O #M4 (250000)
M4 M4GOOD TM4I1 FM4I2 FM4O #M4 (250000)
M4 M4GOOD FM4I1 TM4I2 FM4O #M4 (250000)
M4 M4GOOD FM4I1 FM4I2 FM4O #M4 (250000)
M4 M4BAD TM4I1 TM4I2 FM4O #M4 (24)
M4 M4BAD TM4I1 FM4I2 FM4O #M4 (24)
M4 M4BAD FM4I1 TM4I2 FM4O #M4 (24)
M4 M4BAD FM4I1 FM4I2 FM4O #M4 (24)
M4 M4BAD TM4I1 TM4I2 TM4O #M4 (1)
M4 M4BAD TM4I1 FM4I2 TM4O #M4 (1)
M4 M4BAD FM4I1 TM4I2 TM4O #M4 (1)
M4 M4BAD FM4I1 FM4I2 TM4O #M4 (1)
M5 M5GOOD TM5I1 TM5I2 TM5O #M5 (250000)
M5 M5GOOD TM5I1 FM5I2 FM5O #M5 (250000)
M5 M5GOOD FM5I1 TM5I2 FM5O #M5 (250000)
M5 M5GOOD FM5I1 FM5I2 FM5O #M5 (250000)
M5 M5BAD TM5I1 TM5I2 FM5O #M5 (24)
M5 M5BAD TM5I1 FM5I2 FM5O #M5 (24)
M5 M5BAD FM5I1 TM5I2 FM5O #M5 (24)
M5 M5BAD FM5I1 FM5I2 FM5O #M5 (24)
M5 M5BAD TM5I1 TM5I2 TM5O #M5 (1)
M5 M5BAD TM5I1 FM5I2 TM5O #M5 (1)
M5 M5BAD FM5I1 TM5I2 TM5O #M5 (1)
M5 M5BAD FM5I1 FM5I2 TM5O #M5 (1)
frame M1 #M1 M2 #M2 M3 #M3 M4 #M4 M5 #M5 #frame (1)
```

Figure 27. A set of SP patterns modelling input-output relations in the electronic circuit shown in Figure 25. They were supplied as Old patterns to the SP model for the building of the multiple alignment shown in Figure 28. *Key*: "T" = true (information is correct); "F" = false (information is incorrect); "M1", "M2", "M3", "M4", "M5" = components of the circuit; "GOOD", "BAD" indicates whether a component is good or bad; "I1", "I2" = first and second inputs of a component; "O" = Output of a component.

```
0         1         2         3         4         5         6

                    frame
                    M1 -------------------------- M1
                                                  M1BAD
TM1I1 ------------------------------------------- TM1I1
TM1I2 ------------------------------------------- TM1I2
                                                  FM1O
                    #M1 ------------------------- #M1
                    M2 ------------------------------------ M2
                                                            M2GOOD
TM2I1 ----------------------------------------------------- TM2I1
TM2I2 ----------------------------------------------------- TM2I2
                                                            TM2O
                    #M2 ----------------------------------- #M2
                    M3 ------ M3
                              M3GOOD
TM3I1 ----------------------- TM3I1
TM3I2 ----------------------- TM3I2
                              TM3O
                    #M3 ----- #M3
                    M4 ---------------- M4
                                        M4GOOD
                                        TM4I1 ------------- TM4I1
                                        FM4I2 --- FM4I2
FM4O ---------------------------------- FM4O
                    #M4 --------------- #M4
          M5 ------ M5
          M5GOOD
          TM5I1 ------------- TM5I1
          TM5I2 ------------------------------------------- TM5I2
TM5O ---- TM5O
          #M5 ----- #M5
                    #frame

0         1         2         3         4         5         6
```

Figure 28. The best multiple alignment found by the SP model with "TM1I1 TM1I2 TM2I1 TM2I2 TM3I1 TM3I2 FM4O TM5O" in New and the patterns shown in Figure 27 in Old.

Table 2. Seven alternative diagnoses of faults in the circuit shown in Figure 25, derived from multiple alignments created by the SP model with "TM1I1 TM1I2 TM2I1 TM2I2 TM3I1 TM3I2 FM4O TM5O" in New and the patterns from Figure 27 in Old. The relative probability of each diagnosis is shown in the second column.

Bad Component(s)	*Relative Probability*
M1	0.6664
M4	0.3332
M1, M3	0.00013
M1, M2	0.00013
M1, M4	6.664e-5
M3, M4	6.664e-5
M1, M2, M3	2.666e-8

It is interesting to see that the best diagnosis derived by the SP model (M 1 is bad) and the second best diagnosis (M 4 is bad) are in accordance with first two diagnoses obtained by Pearl's method. The remaining five diagnoses derived by the SP model are different from the one obtained by Pearl's method (M 2 and M 5 are bad), but this is not altogether surprising, because detailed frequencies or probabilities are different from Pearl's example, and there are differences in assumptions that have been made.

11. INFORMATION STORAGE AND RETRIEVAL

The SP theory provides a versatile model for database systems, with the ability to accommodate object-oriented structures, as well as relational "tuples" and network and tree models of data [35]. It lends itself most directly to information retrieval in the manner of query-by-example, but it appears to have potential to support the use of natural language or query languages, such as SQL (Structured Query Language).

Unlike some ordinary database systems:

- The storage and retrieval of information is integrated with other aspects of intelligence, such as pattern recognition, reasoning, planning, problem solving and learning—as outlined elsewhere in this article.
- The SP system provides a simple but effective means of combining class hierarchies with part-whole hierarchies, with inheritance of attributes (Section 9.1).
- It provides for cross-classification with multiple inheritance.
- There is flexibility and versatility in the representation of knowledge arising from the fact that the system does not distinguish "parts" and "attributes" (Section 4.2.1 in [35]).
- Likewise, the absence of a distinction between "class" and "object" facilitates the representation of knowledge and eliminates the need for a "metaclass" (Section 4.2.2 in [35]).
- SP patterns provide a simpler and more direct means of representing entity-relationship models than do relational tuples (Section 4.2.3 in [35]).

12. PLANNING AND PROBLEM SOLVING

The SP framework provides a means of planning a route between two places, and, with the translation of geometric patterns into textual form, it can solve the kind of geometric analogy problem that may be seen in some puzzle books and IQ tests (*BK*, Chapter 8).

Figure 29 shows an example of the latter kind of problem. The task is to complete the relationship "A is to B as C is to ?" using one of the figures, "D", "E", "F" or "G", in the position marked with "?". For this example, the "correct" answer is clearly "E". Quote marks have been used for the word "correct", because in many problems of this type, there may be two or even more alternative answers for which cases can be made, and there is a corresponding uncertainty about which answer is the right one.

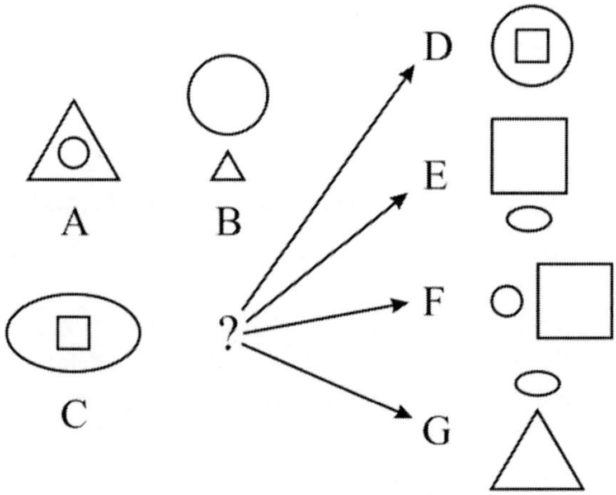

Figure 29. A geometric analogy problem.

Computer-based methods for solving this kind of problem have existed for some time (e.g., Evans' well-known heuristic algorithm [62]). In more recent work [63,64], minimum length encoding principles have been applied to good effect. This kind of problem may also be understood in terms of the SP concepts.

As in most previous work, the proposed solution assumes that some mechanism is available, which can translate the geometric forms in each problem into patterns of text symbols, like other patterns in this article. For example, item "A" in Figure 29 may be described as "small circle inside large triangle".

How this kind of translation may be done is not part of the present proposals (one such translation mechanism is described in [62]). As noted elsewhere [64], successful solutions for this kind of problem require consistency in the way the translation is done. For this example, it would be unhelpful if item "A" in Figure 29 were described as "large triangle outside small circle", while item "C" were described as "small square inside large ellipse". For any one puzzle, the description needs to stick to one or other of "X outside Y" or "Y inside X"—and likewise for "above/below" and "left-of/right-of".

Given that the diagrammatic form of the problem has been translated into patterns as just described, this kind of problem can be cast as a problem of partial matching, well within the scope of the SP model. To do this, symbolic representations of item A and item B inFigure 29 are treated as a single pattern, thus:

```
small circle inside large triangle ;
 large circle above small triangle
```

and this pattern is placed in New. Four other patterns are constructed by pairing a symbolic representation of item C (on the left) with symbolic representations of each of D, E, F and G (on the right), thus:

```
C1 small square inside large ellipse ;
    D small square inside large circle #C1
C2 small square inside large ellipse ;
    E large square above small ellipse #C2
C3 small square inside large ellipse ;
    F small circle left-of large square #C3
C4 small square inside large ellipse ;
    G small ellipse above large rectangle #C4.
```

These four patterns are placed in Old, each with an arbitrary frequency value of 1.

Figure 30 shows the best multiple alignment found by the SP model with New and Old, as just described. The multiple alignment is a partial match between the New pattern (in column 0) and the second of the four patterns in Old (in column 1). This corresponds with the "correct" result (item E), as noted above.

13. COMPRESSION OF INFORMATION

Since information compression is central to the workings of the SP system, it is natural to consider whether the system might provide useful insights in that area. In that connection, the most promising aspects of the SP system appear to be:

- The discovery of recurrent patterns in data via the building of multiple alignments, with heuristic search to sift out the patterns that are most useful in terms of compression.

- The potential of the system to detect and encode discontinuous dependencies in data. It appears that there is potential here to extract kinds of redundancy in information that are not accessible via standard methods for the compression of information.

```
0           1

            C2
small ----  small
circle      square
inside ---  inside
large ----  large
triangle    ellipse
; -------- ;
            E
large ----  large
circle      square
above ----  above
small ----  small
triangle    ellipse
            #C2

0           1
```

Figure 30. The best multiple alignment found by the SP model for the patterns in New and Old as described in the text.

In terms of the trade-off that exists between computational resources that are required and the level of compression that can be achieved, it is intended that the system will operate towards the "up market" end of the spectrum—by contrast with LZW algorithms and the like, which have been designed to be "quick-and-dirty", sacrificing performance for speed on low-powered computers.

14. HUMAN PERCEPTION, COGNITION AND NEUROSCIENCE

Since much of the inspiration for the SP theory has come from evidence, mentioned in Section 2.1, that, to a large extent, the workings of brains and nervous systems may be understood in terms of information compression, the theory is about perception and cognition, as well as artificial intelligence and mainstream computing.

That said, the main elements of the theory—the multiple alignment concept in particular—are theoretical constructs derived from what appears to be necessary to model, in an economical way, such things as pattern recognition, reasoning, and so on. In *BK* (Chapter 12), there is some discussion of how the SP concepts relate to a selection of issues in human perception and cognition. A particular interest at the time of writing (after that chapter was written) is the way that the SP theory may provide an alternative to quantum probability as an explanation of phenomena, such as the "conjunction fallacy" (see, for example [65]).

In *BK* (Chapter 11), I have described in outline, and tentatively, how such things as SP patterns and multiple alignments may be realised with neurons and connections between them. The cortex of the brains of mammals—which is, topologically, a two-dimensional sheet—may be, in some respects, like a sheet of paper on which *pattern assemblies* may be written. These are neural analogues of SP patterns, shown schematically in Figure 31. Unlike information written on a sheet of paper, there are neural connections between patterns—as shown in the figure—and communications amongst them.

These proposals, which are adapted with modifications from Hebb's [66] concept of a "cell assembly", are very different from how artificial "neural networks" are generally conceived in computer science [67].

As noted in Section 5.4, learning in the SP system is very different from learning in that kind of network, or Hebbian learning.

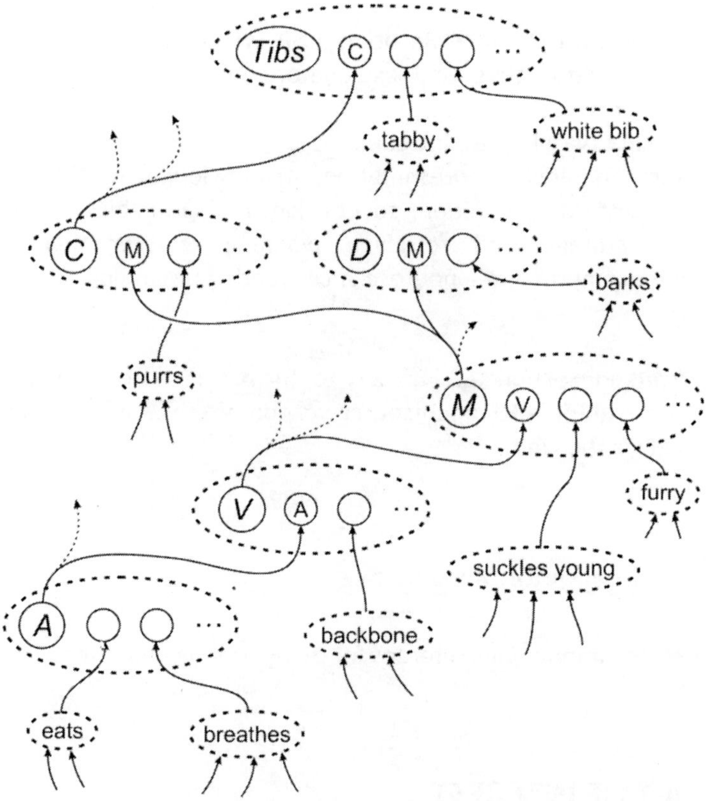

Figure 31. Schematic representation of hypothesised neural analogues of SP patterns and their inter-connections. *Key*: "C" = cat, "D" = dog, "M" = mammal, "V" = vertebrate, "A" = animal, "..." = further structure that would be shown in a more comprehensive example. Pattern assemblies are surrounded by broken lines, and each neuron is represented by an unbroken circle or ellipse. Lines with arrows show connections between pattern assemblies and the flow of sensory signals. Connections between neurons within each pattern assembly are not marked.

15. CONCLUSIONS

The SP theory aims to simplify and integrate concepts across artificial intelligence, mainstream computing and human perception and cognition, with information compression as a unifying theme. The matching and unification of patterns and the concept of multiple alignment are central ideas.

In accordance with Occam's Razor, the SP system combines conceptual simplicity with descriptive and explanatory power. A relatively simple mechanism provides an interpretation for a range of concepts and phenomena in several areas, including concepts of "computing", aspects of mathematics and logic, representation of knowledge, natural language processing, pattern recognition, several kinds of probabilistic reasoning, information storage and retrieval, planning and problem solving, information compression, neuroscience, and human perception and cognition.

As suggested in Section 3.2, an aid to further research would be the creation of a high-parallel, open-source version of the SP machine that may be accessed via the web.

ACKNOWLEDGEMENTS

I am grateful to anonymous referees for many useful comments.

CONFLICT OF INTEREST

The author declares no conflict of interest.

REFERENCES AND NOTES

1. Apart from the period between early 2006 and late 2012, when I was working on other things.

2. See www.cognitionresearch.org/sp.htm#PUBS.

3. Wolff, J.G. *Unifying Computing and Cognition: the SP Theory and Its Applications*; CognitionResearch.org.uk: Menai Bridge, UK, 2006.

4. Some of the text and figures in this article come from the book, with permission. Details of other permissions are given at appropriate points in the article.

5. Attneave, F. Some informational aspects of visual perception. *Psychol. Rev.* **1954**, *61*, 183–193.

6. Barlow, H.B. Sensory Mechanisms, the Reduction of Redundancy, and Intelligence. In *The Mechanisation of Thought Processes*; Her Majesty's Stationery Office: London, UK, 1959; pp. 535–559.

7. Barlow, H.B. Trigger Features, Adaptation and Economy of Impulses. In *Information Processes in the Nervous System*; Leibovic, K.N., Ed.; Springer: New York, NY, USA, 1969; pp. 209–230.

8. Also relevant and still of interest is Zipf's [68] *Human Behaviour and the Principle of Least Effort*. Incidentally, Barlow later suggested that "... the [original] idea was right in drawing attention to the importance of redundancy in sensory messages ... but it was wrong in emphasizing the main technical use for redundancy, which is compressive coding." (p. 242 in [69]). As we shall see, the SP theory is closer to Barlow's original thinking than what he said later.

9. This focus on compression of information in binocular vision is distinct from the more usual interest in the way that slight differences between the two images enables us to see the scene in depth.

10. Wolff, J.G. Learning Syntax and Meanings through Optimization and Distributional Analysis. In *Categories and Processes in Language Acquisition*; Levy, Y., Schlesinger, I.M., Braine, M.D.S., Eds.; Lawrence Erlbaum: Hillsdale, NJ, USA, 1988; pp. 179–215.

11. See www.cognitionresearch.org/lang learn.html.

12. Solomonoff, R.J. A formal theory of inductive inference. Part I. *Inf. Control* **1964**, *7*, 1–22.

13. Solomonoff, R.J. A formal theory of inductive inference. Part II. *Inf. Control* **1964**, *7*, 224–254.

14. Li, M.; Vitänyi, P. *An Introduction to Kolmogorov Complexity and Its Applications*; Springer: New York, NY, USA, 2009.

15. Newell, A. You can't Play 20 Questions with Nature and Win: Projective Comments on the Papers in This Symposium. In *Visual Information Processing*; Chase, W.G., Ed.; Academic Press: New York, NY, USA, 1973; pp. 283–308.

16. Laird, J.E. *The Soar Cognitive Architecture*; MIT Press: Cambridge, MA, USA, 2012.

17. Anderson, J.R.; Bothell, D.; Byrne, M.D.; Douglass, S.; Lebiere, C.; Qin, Y. An integrated theory of the mind. *Psychol. Rev.***2004**, *111*, 1036–1060.

18. Schmidhuber, J., Thörisson, K.R., Looks, M., Eds.; *Artificial General Intelligence: 4th International Conference, AGI 2011, Mountain View, CA, USA, August 3–6, 2011, Proceedings*; Volume 6830, Lecture Notes in Artificial Intelligence; Springer: New York, NY, USA, 2011.

19. Dodig-Crnkovic, G. Significance of models of computation, from Turing model to natural computation. *Minds Mach.* **2011**, *21*, 301–322.

20. Steunebrink, B.R.; Schmidhuber, J. A family of Gödel machine implementations. In [18]. Available online: www.idsia.ch/ juergen/ agi2011bas.pdf (accessed on 31 July 2013).

21. Hutter, M. *Universal Artificial Intelligence: Sequential Decisions based on Algorithmic Probability*; Springer: Berlin, Germany, 2005.

22. Wolff, J.G. Simplicity and power—some unifying ideas in computing. *Comput. J.* **1990**, *33*, 518–534.

23. Of course, people can and do learn with assistance from teachers and others. However, unsupervised learning has been a focus of interest in developing the SP theory, since it is clear that much of our learning is done without assistance and because unsupervised learning raises some interesting issues and yields some useful insights, as outlined in Section 5.2.

24. The source code for the models, with associated documents and files, may be downloaded via links under the heading "SOURCE CODE" at the bottom of the page on http://bit.ly/WtXa3g (accessed on 5 August 2013)..

25. As in ordinary search engines and, indeed, in the brains of people and other animals, high levels of parallelism are needed to achieve speedy processing with large data sets (see also Section 4.3 and Section 5.1.5.).

26. Wolff, J.G. The SP theory of intelligence: Benefits and applications. 2013; in preparation. Available online: http://bit.ly/12YmQJW (accessed on 31 July 2013).

27. Wolff, J.G. Application of the SP theory of intelligence to the understanding of natural vision and the development of computer vision. 2013; in preparation. Available online: http://bit.ly/Xj3nDY (accessed on 31 July 2013).

28. Prince, S.J.D. *Computer Vision: Models, Learning, and Inference*; Cambridge University Press: Cambridge, UK, 2012.

29. Whether multiple alignments are shown with patterns in rows or in columns depends largely on what fits best on the page.

30. Cover, T.M.; Thomas, J.A. *Elements of Information Theory*; Wiley: New York, NY, USA, 1991.

31. Although this scheme is slightly less efficient than the well-known Huffman scheme, it has been adopted, because, unlike the Huffman scheme, it does not produce anomalous results when probabilities are derived from code sizes, as described in *BK*(Section 3.7).

32. See, for example, "Sequence alignment", Wikipedia. Available online: en.wikipedia.org/wiki/Sequence alignment ((accessed on 8 May 2013).

33. Sankoff, D.; Kruskall, J.B. *Time Warps, String Edits, and Macromolecules: The Theory and Practice of Sequence Comparisons*; Addison-Wesley: Reading, MA, USA, 1983.

34. *WinMerge*, latest stable version 2.14.0; Open Source differencing and merging tool for Windows. Available online: http://winmerge.org (accessed on 31 July 2013).

35. Wolff, J.G. Towards an intelligent database system founded on the SP theory of computing and cognition. *Data Knowl. Eng.***2007**, *60*, 596–624.

36. Dorigo, M.; Gambardella, L.M. Ant colony system: A cooperative learning approach to the traveling salesman problem. *IEEE Trans. Evol. Comput.* **1997**, *1*, 53–66.

37. Thus "computing as compression" does not imply that all redundancy is bad and should be removed. Redundancy in information is often useful in, for example, understanding speech in noisy conditions (*cf.*, Section 4.2.2) or in backup copies for data.

38. Wolff, J.G. Language acquisition, data compression and generalization. *Lang. Commun.* **1982**, *2*, 57–89.

39. In this and other examples in this subsection, we shall assume that letters are analogues of low-level perceptual features in speech, such as formant ratios or formant transitions.

40. Wolff, J.G. The discovery of segments in natural language. *Br. J. Psychol.* **1977**, *68*, 97–106.

41. Wolff, J.G. Language acquisition and the discovery of phrase structure. *Lang. Speech* **1980**, *23*, 255–269.

42. The Chomskian doctrine that children are born with a knowledge of "universal grammar" fails to account for the specifics of syntactic forms in different languages, and it depends on the still-unproven idea that there is something of substance that is shared by all the world's languages.

43. Relevant evidence comes from cases where children learn to understand language even though they have little or no ability to speak [70,71]—so that there is little or nothing for anyone to correct.

44. If an error is not rare, it is likely to acquire the status of a dialect or idiolect variation and cease to be regarded as an error.

45. Such as: learning in the kinds of artificial neural network that are popular in computer science; Hebb's [66] concept of learning; Pavlovian learning; and Skinnerian learning.

46. Turing, A.M. On computable numbers, with an application to the Entscheidungsproblem. *Proc. Lond. Math. Soc.* **1936**, *42*, 230–265.

47. Turing, A.M. On computable numbers, with an application to the Entscheidungsproblem: a correction. *Proc. Lond. Math. Soc.***1937**, *43*, 544–546.

48. Post, E.L. Formal reductions of the general combinatorial decision problem. *Am. J. Math.* **1943**, *65*, 197–268.

49. Chaitin, G.J. Randomness in arithmetic. *Sci. Am.* **1988**, *259*, 80–85.

50. Wolff, J.G. The SP Theory and the Representation and Processing of Knowledge. In *Soft Computing in Ontologies and Semantic Web*; Ma, Z., Ed.; Springer-Verlag: Heidelberg, Germany, 2006; pp. 79–101.

51. Wolff, J.G. Medical diagnosis as pattern recognition in a framework of information compression by multiple alignment, unification and search. *Decis. Support Syst.* **2006**, *42*, 608–625.

52. Iwanska, L.; Zadrozny, W. Introduction to special issue on context in natural language processing. *Comput. Intell.* **1997**, *13*, 301–308.

53. Chomsky, N. *Syntactic Structures*; Mouton: The Hague, The Netherlands, 1957.

54. Pereira, F.C.N.; Warren, D.H.D. Definite clause grammars for language analysis—a survey of the formalism and a comparison with augmented transition networks. *Artif. Intell.* **1980**, *13*, 231–278.

55. In this figure, the sentence, "it is wash ed", could have been represented more elegantly as, "i t i s w a s h e d", as in previous examples. The form shown here has been adopted, because it helps to stop multiple alignments growing too large. Likewise, with Figure 14.

56. Oliva, A.; Torralba, A. The role of context in object recognition. *Trends Cogn. Sci.* **2007**, *11*, 520–527.

57. Although the term "heterarchy" is not widely used, it can be useful as a means of referring to hierarchies in which, as in the example in the text, a given node may appear in two or more higherlevel nodes that are not themselves hierarchically related. In the SP framework, there may be heterarchies in both class-inclusion structures and part-whole structures. However, to avoid the clumsy expression "hierarchy or heterarchy", the term "hierarchy" is used in most parts of this article as a shorthand for both concepts.

58. Pothos, E.M.; Wolff, J.G. The Simplicity and Power model for inductive inference. *Artif. Intell. Rev.* **2006**, *26*, 211–225.

59. Pearl, J. *Probabilistic Reasoning in Intelligent Systems: Networks of Plausible Inference*, revised second printing ed.; Morgan Kaufmann: San Francisco, CA, USA, 1997.

60. Likewise, a travel booking clerk using a database of all flights between cities will assume that, if no flight is shown between, say, Edinburgh and Paris, then no such flight exists. In systems like Prolog, the closed-world assumption is the basis of "negation as failure": If a proposition cannot be proven with the clauses provided in a Prolog program, then, in terms of that store of knowledge, the proposition is assumed to be false.

61. Some of the frequencies shown in Figure 22 are intended to reflect the two probabilities suggested for this example in [59] (p. 49): "... the [alarm] is sensitive to earthquakes and can be accidentally (p = 0:20) triggered by one. ... if an earthquake had occurred, it surely (p = 0:40) would be on the [radio] news."

62. Evans, T.G. A program for the solution of a class of geometric-analogy intelligence-test questions. In *Semantic Information Processing*; Minsky, M.L., Ed.; MIT Press: Cambridge, MA, USA, 1968; pp. 271–353.

63. Belloti, T.; Gammerman, A. Experiments in solving analogy problems using Minimal Length Encoding. *Appl. Decis. Technol.***1996**, *95*, 209–220.

64. Gammerman, A.J. The representation and manipulation of the algorithmic probability measure for problem solving. *Ann. Math. Artif. Intell.* **1991**, *4*, 281–300.

65. Pothos, E.M.; Busemeyer, J.R. Can quantum probability provide a new direction for cognitive modeling? *Behav. Brain Sci.* **2013**,*36*, 255–327.

66. Hebb, D.O. *The Organization of Behaviour*; John Wiley & Sons: New York, NY, USA, 1949.

67. See, for example, "Artificial neural network". *Wikipedia*, Available online: http://winmerge.org (accessed on 31 July 2013).

68. Zipf, G.K. *Human Behaviour and the Principle of Least Effort*; Hafner: New York, NY, USA, 1949.

69. Barlow, H.B. Redundancy reduction revisited. *Netw. Comput. Neural Syst.* **2001**, *12*, 241–253.

70. Lenneberg, E.H. Understanding language without the ability to speak. *J. Abnorm. Soc. Psychol.* **1962**, *65*, 419–425.

71. Brown, R. *A First Language: The Early Stages*; Penguin: Harmondsworth, UK, 1973.

Chapter 4

The SP Theory of Intelligence: Benefits and Applications

J Gerard Wolff

CognitionResearch.org, Menai Bridge LL59 5LR, UK;

ABSTRACT

This article describes existing and expected benefits of the *SP theory of intelligence*, and some potential applications. The theory aims to simplify and integrate ideas across artificial intelligence, mainstream computing, and human perception and cognition, with information compression as a unifying theme. It combines conceptual simplicity with descriptive and explanatory power across several areas of computing and cognition. In the *SP machine*—an expression of the SP theory which is currently realized in the form of a computer model—there is potential for an overall simplification of computing systems, including software. The SP theory promises deeper insights and better solutions in several areas of application including, most notably, unsupervised learning, natural language processing, autonomous robots, computer vision, intelligent databases, software engineering, information compression, medical diagnosis and big data. There is also potential in areas such as the semantic web, bioinformatics, structuring of documents, the detection of computer viruses, data fusion, new kinds of computer, and the development of scientific theories. The theory promises seamless integration of structures and functions within and between different areas of application. The potential value, worldwide, of these benefits and applications is at least

$190 billion each year. Further development would be facilitated by the creation of a high-parallel, open-source version of the SP machine, available to researchers everywhere.

KEYWORDS

artificial intelligence; information compression; unsupervised learning; natural language processing; pattern recognition

1. INTRODUCTION

The *SP theory of intelligence* aims to simplify and integrate concepts across artificial intelligence, mainstream computing and human perception and cognition, with information compression as a unifying theme. This article describes existing and expected benefits of the SP theory and some of its potential applications.

The theory is described most fully in [1] and more briefly in an extended overview [2]. This article should be read in conjunction with either or both of those accounts.

In brief, the existing and expected benefits of the theory are:

- Conceptual simplicity combined with descriptive and explanatory power.
- Simplification of computing systems, including software.
- Deeper insights and better solutions in several areas of application.
- Seamless integration of structures and functions within and between different areas of application.

These points will be discussed, each in its own section, below. But first, as a background and context to what follows, there is a little philosophy, and an introduction to the SP theory with some associated ideas.

2. A LITTLE PHILOSOPHY

Amongst alternative criteria for the success of a theory, what appears to be the most widely accepted is the principle ("Occam's Razor") that a good theory should combine *simplicity* with descriptive or explanatory *power* [3]. This equates with the idea—which echoes the underlying theme of the SP theory itself—that a good theory should compress empirical data via a reduction of "redundancy" in the data (thus increasing its "simplicity"), whilst retaining as much as possible of its non-redundant descriptive "power" [4]. As John Barrow has written: "Science is, at root, just the search for compression in the world." [5] (p. 247).

These principles are prominent in most areas of science: the Copernican heliocentric theory of the Sun and the planets is rightly seen to provide a simpler and more powerful account of the facts than Ptolemy's Earth-centered theory, with its complex epicycles; physicists have a keen interest in the possibility that quantum mechanics and relativity might be unified; biology would be greatly impoverished without modern understandings of evolution, genetics, and DNA; and so on. But in research in computer science, including artificial intelligence, there is a fragmentation of the field into a myriad of concepts and many specialisms, with little in the way of an overarching theory to pull everything together [6].

Perhaps the concept of a universal Turing machine [7,8] provides what is needed? That theory has of course been brilliantly successful—a basis for the current generation of computers and the many useful things that they can do. But it does not solve the problem of fragmentation and, although Alan Turing recognized that computers might become intelligent [9], the Turing theory, in itself, does not tell us how!

Are we getting closer to that goal? Yes, certainly. But as Hubert Dreyfus has suggested ([10] (p. 12) and [11]), we may be getting closer to achieving real intelligence in computers in the same sense that a man climbing a tree is getting closer to reaching the moon.

Whether or not the SP theory succeeds in getting us to the AI moon, I believe there is a pressing need, in computer science and AI, for a stronger focus on the simplification and integration of ideas. In all areas of application, including quite specialized areas, a theory that scores well in terms of simplicity and power is, compared with any weaker theory, likely to yield deeper insights, better solutions, and better integration of structures and functions, both within a given area and amongst different areas (see also Section 6 and Section 7).

3. THE SP THEORY AND THE SP MACHINE: A SUMMARY

In broad terms, the SP theory has three main elements:

- All kinds of knowledge are represented with *patterns*: arrays of atomic symbols in one or two dimensions.
- At the heart of the system is compression of information via the matching and unification (merging) of patterns, and the building of *multiple alignments* like the one shown in Figure 1.
- The system learns by compressing "New" patterns to create "Old" patterns like those shown in rows 1 to 8 in Figure 1.

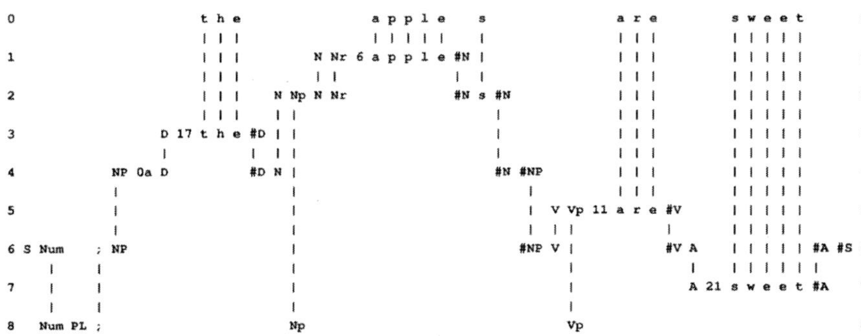

Figure 1. The best multiple alignment created by the SP computer model with a store of Old patterns like those in rows 1 to 8 (representing grammatical structures, including words) and a New pattern (representing a sentence to be parsed) shown in row 0.

The system is realized in the form of a computer model [12]. It is envisaged that this will be developed into a high-parallel, open-source *SP machine*, an expression of the SP theory and a means for it to be applied (Section 3.2 in [2]). It is envisaged that this will be a facility for researchers everywhere to investigate what can be done with the system and create new versions of it. How things may develop is shown schematically in Figure 2.

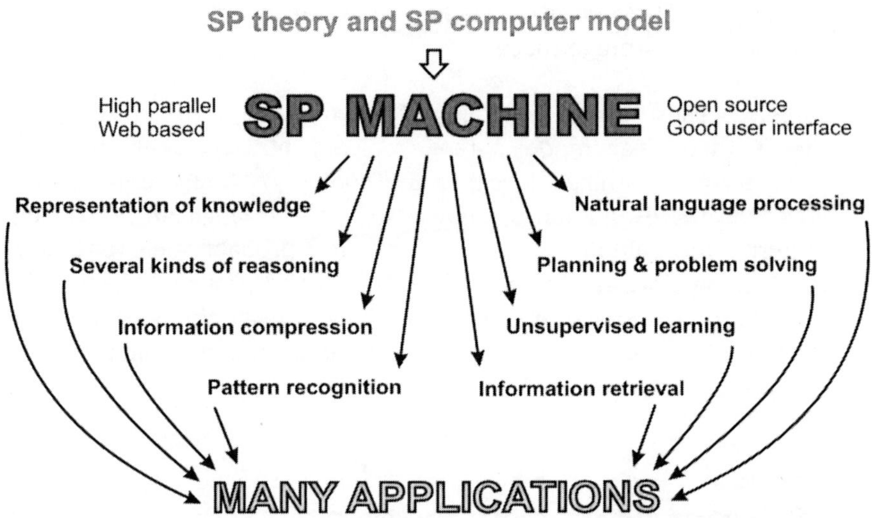

Figure 2. Schematic representation of the development and application of the proposed SP machine. Reproduced with permission from Figure 2 in [2].

4. COMBINING CONCEPTUAL SIMPLICITY WITH DESCRIPTIVE AND EXPLANATORY POWER

In principle, it should be possible to evaluate scientific theories quite precisely in terms of conceptual simplicity and descriptive and explanatory power—and to compare one theory with another—by measuring their ability to compress empirical data. But, for much of science, techniques and technologies are not yet well enough developed to make this feasible, and there is in any case the difficulty that it is rare for rival theories to address precisely the same body of empirical data. So in evaluating theories we normally need to rely on more-or-less informal judgments of simplicity and power (see also Section 6.10.7).

In those terms, the SP theory appears to score well:

- The SP theory, including the multiple alignment concept, is not trivially simple but it is not unduly complicated either. The SP computer model (Sections 3.9, 3.10 and 9.2 in [1]; Section 3.1 in [2]), which is, apart from some associated thinking, the most comprehensive expression of the theory as it stands now, is embodied in an "exec" file requiring less than 500 KB of storage space.

- Largely because of the versatility of the multiple alignment concept, the SP theory can model a range of concepts and phenomena in: unsupervised learning, concepts of "computing", mathematics and logic, the representation of knowledge, natural language processing, pattern recognition, several kinds of probabilistic reasoning, information storage and retrieval, planning and problem solving, and information compression; and it has things to say about concepts in neuroscience and in human perception and cognition. (Chapters 4–13 in [1]; Sections 5–13 in [2]).

5. SIMPLIFICATION OF COMPUTING SYSTEMS, INCLUDING SOFTWARE

Apart from the simplification and integration of concepts in artificial intelligence, mainstream computing, and human perception and cognition (Section 4), the SP theory can help to simplify computing systems, including software.

The principle to be described here is already familiar in the way databases and expert systems are structured.

Early databases were each written as a monolithic system containing everything needed for its operation, and likewise for early expert systems. But it soon became apparent that, in the case of databases, a lot of effort could be saved by creating a generalized "database management system" (DBMS), with a user interface and mechanisms for the storage and retrieval of information, and then creating new databases by loading the DBMS with different bodies of information, according to need. In a similar way, an "expert system shell", with a user interface and with mechanisms for inference and for the storage and retrieval of information, eliminates the need to create those facilities repeatedly in each new expert system.

The SP system takes this principle further. It aims to provide a general-purpose "intelligence"—chiefly the multiple alignment framework—and thus save the need to create those kinds of mechanisms repeatedly in different AI applications: pattern recognition, natural language processing, several kinds of reasoning, planning, problem solving, unsupervised learning, and more.

The principle extends to conventional applications, since the matching and unification of patterns and a process of searching amongst alternative matches for patterns (which are elements of the multiple alignment framework), are used in many kinds of application outside the world of artificial intelligence.

The way in which this principle may help to simplify computing systems is shown schematically in Figure 3. In a conventional computer, shown at the top of the figure, there is a central processing unit (CPU) which is relatively simple and without "intelligence". Software in the system, shown to the right, is a combination of two things:

- Domain-specific knowledge such as knowledge of accountancy, geography, the organization and procedures of a business, and so on [13].

- Processing instructions to provide the intelligence that is missing in the CPU, chiefly processes for the matching and unification of patterns, and for searching amongst alternative matches to find one or more that are "good". These kinds of processing instructions—let us call them MUP instructions—are used in many different kinds of application, meaning that there is a considerable amount of redundancy in conventional computing systems—where the term "system" includes both the variety of programs that may be run, as well as the hardware on which they run.

In the SP machine, shown schematically at the bottom of the figure, the "CPU" [14] is more complicated (as suggested by the larger size of "CPU" in the figure) and aims to provide much of the "intelligence" that is missing in the CPU of a conventional computer. This should mean that MUP instructions, and their many repetitions, may be largely eliminated from the software, leaving just the domain-specific knowledge. The increased complexity of the "CPU" should be more than offset by cutting out much of the redundancy in the software, meaning an overall reduction in complexity—as suggested by the relatively small size of the figure that represents the SP machine, compared with the figure that represents a conventional computer.

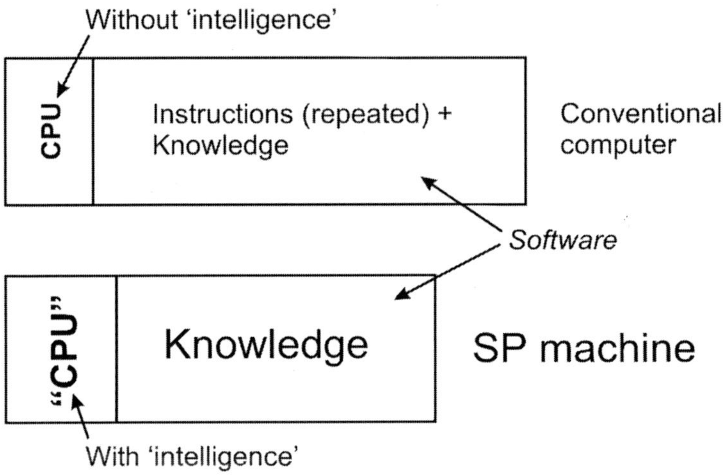

Figure 3. Schematic representations of a conventional computer and an SP machine, as discussed in the text. Adapted with permission from Figure 4.7 in [1].

An overall simplification of computing systems can bring benefits like these:

- *Savings in development effort and associated costs*. With more intelligence in the CPU there should be less need for it to be encoded in applications.

- *Savings in development time*. With a reduced need for hand crafting, applications may be developed more quickly.

- *Savings in storage costs*. There may be useful economies in the storage space required for application code.

- Related benefits noted in Section 6.7.

6. DEEPER INSIGHTS AND BETTER SOLUTIONS IN SEVERAL AREAS OF APPLICATION

This section describes some potential areas of application for the SP theory, and its potential benefits, drawing on the capabilities of the SP system (described in [1] (Chapters 4–11) and [2] (Sections 5–13)). These examples are probably just the tip of the iceberg. To mix our metaphors, there is a rich vein to be mined.

Because of its importance, I believe it is worth repeating the principle, mentioned in Section 2, that, even in specialized areas, a theory like the SP theory that has a broad base of support is likely to provide deeper insights than others, better solutions to problems, and (Section 7) better integration of structures and functions within and between different areas of application. Putative examples from the SP theory include:

- A new method for encoding discontinuous dependencies in syntax (Chapter 5 in [1]; Section 8.1 in [2]), which is arguably simpler and more versatile than older methods.

- The multiple alignment framework, with the use of one simple format for all kinds of knowledge, is likely to facilitate the seamless integration of the syntax and semantics of natural languages, with the kinds of subtle interaction described in Section 6.2, below.

- The multiple alignment framework provides for the seamless integration of class inclusion hierarchies and part-whole hierarchies (Section 6.4 in [1]; Section 9.1 in [2]), something which is missing in most object-oriented systems.

- The use of one framework for both the analysis and production of knowledge (Section 3.8 in [1]; Section 4.5 in [2]) is a significant simplification compared with systems in which there are separate mechanisms.

- The multiple alignment framework provides for the integration of the relational, object-oriented, and network models for databases, with improvements in the relational model [15].

- The multiple alignment framework, with the use of one simple format for all kinds of knowledge, is likely to facilitate the learning of concepts that include diverse kinds of knowledge.

Theories with a narrow focus may work reasonably well in the target area but are likely to fail when the scope is widened. Even if existing technologies in a given area are doing well, there is likely to be a case for examining how the SP theory may be applied, and its potential benefits.

These kinds of benefits of a good theory have been seen repeatedly in the history of science: the germ theory of disease serves us better than the idea that diseases are caused by "bad air"; recognizing the role of oxygen in combustion is more practical than the phlogiston theory; space travel does not work unless we understand the laws of physics; and so on.

6.1. Unsupervised Learning

The first area of application to be considered—unsupervised learning—is a bit different from the others because it is less an area of application in its own right and more a contribution to others, most notably autonomous robots (Section 6.3), computer vision (Section 6.4), software engineering (Section 6.6), big data (Section 6.9), the semantic web (Section 6.10.1), bioinformatics (Section 6.10.2), and the development of scientific theories (Section 6.10.7).

Research developing computer models of language learning (reviewed in [16]; see also [17])—in particular, the importance of information compression in that area—has been one of the main sources of inspiration for the SP theory. But radical restructuring has been needed to meet the goal of simplifying and integrating ideas across artificial intelligence, mainstream computing, and human perception and cognition.

The new model, SP70 [2] (Section 3.1), has already demonstrated an ability to discover generative grammars from samples of artificial languages, including segmental structure, classes of structure, and abstract patterns. But some further work is needed to realize its full potential for learning, as outlined in [2] (Section 3.3). I believe the residual problems are soluble, and that, with their solution, the SP machine will have considerable potential in the areas mentioned above, and others.

A particular area of interest is, of course, the learning of natural languages, perhaps starting with the learning of syntactic structures. On longer time scales, their is potential for the learning of non-syntactic "semantic" structures and their integration with syntax.

6.2. Applications in the Processing of Natural Language

As noted in [2] (Section 8), the SP system supports the parsing and production of language, it provides a simple but effective means of representing discontinuous dependencies in syntax, it is robust in the face of errors in data, it facilitates the integration of syntax with semantics, and it provides a role for context in the processing of language. There are many potential applications, including such things as checking for grammatical errors in text; predictive typing that takes account of grammar as well as the spellings of individual words; natural-language front-ends to databases and expert systems; and the understanding and translation of natural languages.

It is true that some of these things are already done quite well with existing technologies but, as suggested in Section 2,Section 6, and Section 7, there may nevertheless be advantages in exploring what can be done with SP system. For example, there are some subtle interactions between syntax and semantics which which may be difficult to handle without seamless integration of the two:

- Compare "Fafnir, who plays third base on the Little League team, is a fine boy." with "They called their son Fafnir, which is a ridiculous name." [18] (p. 139). The choice of "who" or "which" in these two sentences—a syntactic issue—depends on the *meaning* of "Fafnir". In the first sentence, it means a person, while in the second sentence it means a name.

- As a contrasting example, compare "John gave me the scissors; I am using them now." with "John gave me the two-bladed cutting instrument; I am using it now." [18] (p. 141). In this case, the choice of "them" or "it"—also a syntactic issue—depends on the *syntactic form* of what is referred to: "the scissors" has a plural form while "the two-bladed cutting instrument" has a singular form, although the two expressions have the same meaning.

Apart from the integration of syntax with semantics, the SP system is likely to smooth the path for the integration of natural language processing with other aspects of intelligence: reasoning, pattern recognition, problem solving, and so on.

6.2.1. Towards the Understanding and Translation of Natural Languages

Given the apparent simplicity of representing syntactic and non-syntactic knowledge with patterns [2] (Sections 7 and 8), it may be feasible to use crowd-sourcing in the manner of Wikipedia to create syntactic rules for at least one natural language and a realistically large set of ontologies to provide associated meanings. Existing grammars and existing ontologies may also prove useful in the creation of appropriate sets of SP patterns.

With this kind of knowledge on the web and with a high-parallel, on-line version of the SP machine (as sketched in [2] (Section 3.2)), it should be possible to achieve "understanding" of natural language, either in restricted domains (Section 6.2.2, Section 6.2.3, and Section 6.2.4, below) or more generally.

With the same ontologies and with syntactic knowledge for two or more languages, it may be feasible to achieve translations between languages, with the ontologies functioning as an "interlingua" [19].

The use of an interlingua should yield two main benefits:

- When translations are required amongst several different languages, it is only necessary to create *n* mappings, one between each of the *n* languages and the interlingua. This contrasts with syntax-to-syntax systems, where, in the most comprehensive application of this approach, n!/2(n−2)!n!/2(n-2)! pairs of languages are needed [20].

- It seems likely that the use of an interlingua will be essential if machine translation is ever to reach or exceed the standard that can be achieved by a good human translator.

6.2.2. Natural Language and Information Retrieval

The task of developing a system for understanding natural language may be considerably simplified if we narrow our focus to specific areas: those that require only a relatively restricted range of language forms. Three possibilities are outlined in this and the following two subsections.

On relatively short timescales the SP system may help to overcome some weaknesses in Internet search engines as they are now. For example, if we enter the query "telephone extensions in Anglesey" into any of the leading search engines, we get a mixture of answers: some of them—which are relevant to the query—relate to telephone extensions, while others—which are not relevant—relate to home extensions, building work and the like.

With the SP system, it should be relatively straightforward to overcome that kind of problem, at least for queries about businesses or services in a given area:

- Since a usefully-large range of natural-language questions may be asked with fairly simple language, it would only be necessary—for any given language—to create a relatively simple set of syntactic rules and lexicon, with corresponding semantic structures, all of them expressed with SP patterns.

- It should be possible to create an online tool that would make it easy for the owner or manager of a business to create an ontology for that business that describes what it is, what it does, and the kinds of products or services that it sells. It would of course be necessary for many businesses to provide that kind of information.

- With those two things in place, it should be possible, via the SP system, to create a semantic analysis for a given query (as outlined in [1] (Section 5.7)) and to use that analysis to identify businesses or services that satisfy the query.

With this approach, a query like "telephone extensions in Anglesey" would be parsed into constituents like "telephone extensions" and "in Anglesey", each with a semantic structure. The word "extensions", with a buildings-related meaning, would not figure in those semantic structures, so the unhelpful responses to the query would be eliminated.

6.2.3. Going Beyond FAQs

Many websites provide a list of "frequently asked questions" or FAQs which, as their name suggests, attempt to provide answers to the more commonly-asked questions about the subject of the website. This is OK as far as it goes but it would clearly be more useful if users could simply ask questions in the normal way and receive sensible answers.

To do this with the SP system means developing an ontology for the subject of the website, with an associated vocabulary and syntax, all expressed with SP patterns. This should be feasible to do, partly because of the essential simplicity of representing knowledge in the SP system, and partly because it is likely that the subject matter will be circumscribed.

6.2.4. Interactive Services

In a similar way, it should be possible to develop appropriate ontologies, vocabularies and syntaxes for such applications as booking a hotel room, booking a place on a plane or a train, booking a seat in a restaurant, and so on.

6.3. Towards a Versatile Intelligence for Autonomous Robots

If a robot is to survive and be effective in a challenging environment like Mars, where communication lags restrict the help that can be provided by people, there are advantages if it can be provided with human-like intelligence and versatility—as much as possible. A theory like the SP theory, which provides a unified view of several different aspects of intelligence, is a good candidate for the development of that general intelligence [21].

The SP system has the potential to give robots human-like flexibility and adaptability. Consider, for example, how top-ranking players learn to play games like pool, billiards, or snooker. This is quite different from how some existing robots play pool (see, for example, [22]), where the situation is simplified compared with what a human player faces, and where the robot is provided with a knowledge of geometry and associated mathematics.

For a person to play that kind of game well, they must spend thousands of hours on practice. Potting a ball means visual perception—from the position of the player, not an overhead camera—of the table, the cue ball, the target ball, the pocket and the cue, together with a knowledge of the motor patterns involved in striking the cue ball, and how those motor patterns are shaped by feedback from muscles and from sensory information about the impact between the cue and the cue ball.

Amongst the thousands of such configurations that a player may encounter, he or she must learn which ones lead to success. Getting beyond mere potting of a ball means learning how to impart spin to the cue ball, and judging such things as timings and forces, so that, after a shot, the configuration of balls on the table is favorable for the next shot by the given player, or difficult for his or her opponent. As before, there are thousands of associations to be learned, in this case between visual and motor patterns when the cue ball is struck and the configuration of the balls on the table when they have all stopped moving.

At first sight, the pool-playing robot seems to have the edge over the human player because it does not need all those thousands of hours of practice. But what is missing is that afore-mentioned flexibility and adaptability: without task-specific programming, the robot cannot do such things as make a cup of tea or take the dog for a walk. By contrast, most people can learn to do those kinds of things without difficulty.

In this connection, the main strengths of the SP system are:

- Potential for the kind of visual analysis needed to assimilate the many configurations of balls, pockets, and cue (Section 6.4, next).

- The versatility of the SP framework in the representation and processing of diverse kinds of knowledge (Section 7) should facilitate the seamless integration of visual information about the table, balls, and so on, with information about actions by the player and feedback from muscles and from touch.

- Via its processes for unsupervised learning [2] (Section 5), the SP system should be able to learn associations between visual, motor and feedback patterns, as described above. It appears that similar principles have potential in the kinds of everyday learning mentioned above.

6.4. Computer Vision

The SP system has potential in the area of computer vision, as described in [23]. In brief:

- It has potential to simplify and integrate several areas in computer vision, including feature detection and alignment, segmentation, deriving structure from motion, stitching of images together, stereo correspondence, scene analysis, and object recognition (see, for example, [24]). With regard to the last two topics:

- Since scene analysis is, in some respects, similar to the parsing of natural language, and since the SP system performs well in parsing, it has potential in scene analysis as well. In the same way that a sentence may be parsed successfully without the need for explicit markers of the boundaries between successive words or between successive phrases or clauses, a scene may be analyzed into its component parts without the need for explicit boundaries between objects or other elements in a scene.

- The system provides a powerful and flexible framework for pattern recognition, as outlined in [2] (Section 9), and more fully described in [1] (Chapter 6). It seems likely that this can be generalized for object recognition.

- The SP system has potential for unsupervised learning of the knowledge required for recognition. For example, discrete objects may be identified by the matching and merging of patterns within stereo images (*cf.* [25,26]) or within successive frames in a video, in much the same way that the word structure of natural language may be discovered via the matching and unification of patterns [2] (Section 5.2). The system may also learn such things as classes of entity, and associations between entities, such as the association between black clouds and rain.

- The system is likely to facilitate the seamless integration of vision with other aspects of intelligence: reasoning, planning, problem solving, natural language processing, and so on.

- As noted in [2] (Sections 8 and 9), the system is robust in the face of errors of omission, commission or substitution—an essential feature of any system that is to achieve human-like capabilities in vision.

- With regard to those problems outlined in [2] (Section 3.3), that relate to vision, there are potential solutions:

  - It is likely that the framework can be generalized to accommodate patterns in two dimensions.

- As noted in [2] (Section 3.3), 3D structures may be modeled using 2D patterns, somewhat in the manner of architects' drawings. Knowledge of such structures may be built via the matching and unification of partially-overlapping 2D views [23] (Sections 6.1 and 6.2). Although such knowledge, as with people, may be partial and not geometrically accurate [27], it can, nevertheless, be quite serviceable for such purposes as getting around in one's environment.
- The framework has potential to support both the discovery and recognition of low-level perceptual features.

6.5. A Versatile Model for Intelligent Databases

As described in [15] and outlined in [2] (Section 11), the SP system provides a versatile framework that accommodates several models of data (including object-oriented, relational, network and tree models), that facilitates the integration of different models, and that provides capabilities in reasoning, learning, pattern recognition, problem-solving, and other aspects of intelligence.

An industrial-strength version of the SP machine may prove useful in several areas, including, for example, crime investigations:

- The system would provide a means of storing and managing the data that are gathered in such investigations, often in large amounts.
- It may help in the recognition of features or combinations of features that link a given crime to other crimes, either current or past—and likewise for suspects.
- The system's capabilities in pattern recognition may also serve in the scanning of data to recognize indicators of criminal activity.
- It may prove useful in piecing together coherent patterns from partially-overlapping fragments of information, in much the same way that partially-overlapping digital photographs may be stitched together to create a larger picture.

- Given the capabilities of the SP system for different kinds of reasoning (Chapter 7 in [1] and Section 10 in [2]), including reasoning with information that is not complete, the system may prove useful in suggesting avenues to be explored, perhaps highlighting aspects of an investigation that might otherwise be overlooked.

- Transparency in the representation of knowledge [2] (Section 2.4) and in reasoning via the multiple alignment framework (Chapters 7 and 10 in [1] and Section 10 in [2]) are likely to prove useful where evidence needs to be examined in a court of law, an advantage over "sub-symbolic" systems that yield results without insight.

6.6. Software Engineering

The SP system has potential in software engineering, as described in the following subsections.

6.6.1. Procedural Programming

Procedural programming of a traditional kind is not merely an artifact of the way computers have been designed. It provides an effective means of modeling the kinds of procedures in the real world where the sequence of actions is important. For example, if we are making a cake, we need to mix the ingredients before putting the cake in the oven, and we need to bake the cake before icing it.

It appears that the SP system provides much of what is needed to model real-world procedures of that kind:

- *Procedure*. A sequence of actions may be modeled with a

- one-dimensional SP pattern such as "do_A do_B do_C ...".

- *Variables, values and types*. In the SP framework, any pair of neighboring symbols may function as a slot or variable into which other information may be inserted via the alignment of patterns. For example, the pair of symbols "Num ;" in row 6 of Figure 1 is, in effect, a variable containing the value "PL" from row 8. The "type" of the variable is defined by the range of values it may take, which itself derives from the set of Old patterns provided for the SP model when the multiple alignment was created. In this case, the variable “Num ;” may take the values “PL” (meaning “plural”) or “SNG” (meaning “singular”).

- *Function or subroutine*. In the SP framework, the effect of a function or subroutine may be achieved in much the same way as with variables and values. For example, the pattern “A 21 s w e e t #A” in row 7 of Figure 1 may be seen as a function or subroutine that has been “called” from the higher-level procedure “S Num ; NP #NP V #V A #A #S” in row 6 of the figure.

- *Function with parameters*. As can be seen in Figure 4, the SP system may be run “backwards”, much as can be done with an appropriately-designed Prolog program, generating “data” from “results” (Section 3.8 in [1] and Section 4.5 in [2]). In this example, the pattern “S Num ; NP #NP V #V A #A #S” may be seen as a “function”, and symbols like “PL 0a 17 ...” in row 0 may be seen as “values” for the “parameters” of the function.

- *Conditional statements*. The effect of the values just described is to select elements of the multiple alignment: “PL” selects the pattern “Num PL ; Np Vp”, “0a” selects “NP 0a D #D N #N #NP”, and so on. Each of these selections achieves the effect of a conditional statement in a conventional program: “If ‘PL’ then choose ‘Num PL ; Np Vp’ ”; “If ‘0a’ then choose ‘NP 0a D #D N #N #NP’ ”; and so on.

- *Repetition of procedures*. Conventional programs provide for the repetition of procedures with recursive functions and with statements like *repeat ... until* and *do ... while*. In the SP system, repetition may be encoded with recursive (self-referential) patterns like “X 1 a b c X 1 #X #X” which may be applied as shown in Figure 5.

- *Integration of “programs” and “data”*. Databases provide the mainstay of many software engineering projects and, as described in Chapter 6 in [1], Section 11 in [2] and [15], the SP system promises benefits in that area. In view of the versatility of SP patterns in the multiple alignment framework to represent varied kinds of knowledge [2] (Section 7), there is potential in software engineering projects for the seamless integration of “programs” with “data”.

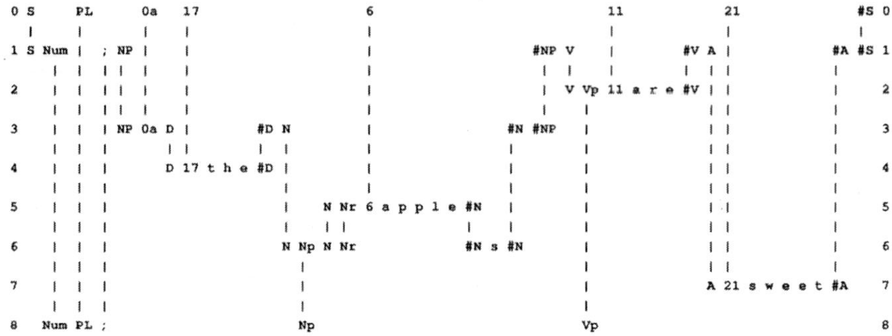

Figure 4. The best multiple alignment created by the SP model with the same Old patterns as were used in the creation of the multiple alignment shown in Figure 1, and a New pattern representing an encoding of a sentence, shown in row 0.

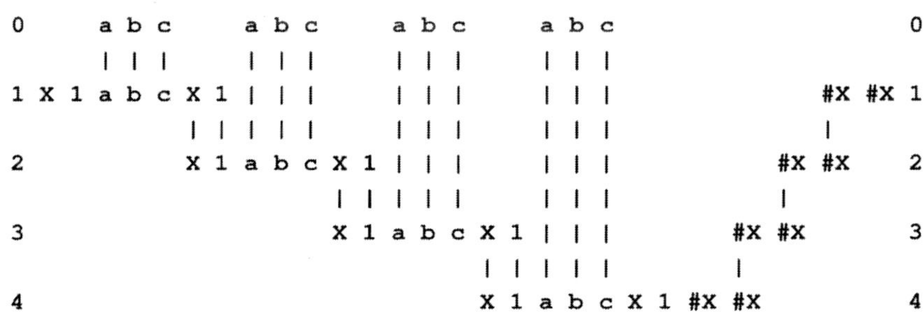

Figure 5. The best multiple alignment produced by the SP model with the New pattern “a b c a b c a b c a b c” and multiple appearances of the Old pattern, “X 1 a b c X 1 #X #X”.

- *Object-oriented design*. The SP system provides for object-oriented design with class hierarchies (including cross-classification), part-whole hierarchies, and inheritance of attributes (Section 6.4 in [1]; Section 9.1 in [2]). In view of the importance of object-oriented design in software engineering, this capability in the SP system is likely to prove useful for applications in that area.

To help smooth the path to working in the SP framework, there may be a case, at least in the early stages, for providing “syntactic sugar” to make things look more like conventional programming.

6.6.2. No Compiling or Interpretation

A potential advantage of the SP system compared with conventional systems is that there would not be any need for compiling or interpretation, meaning the translation of source code into machine code. Any SP program would simply be a set of SP patterns that would be processed directly by the SP machine.

It is true that the current SP computer model is compiled in the conventional way, and the same is likely to be true of any versions of the SP machine that are hosted on a conventional computing system. But the processes of building multiple alignments and creating new knowledge structures does not require compiling or interpretation as normally understood.

6.6.3. Sequential and Parallel Processing

In considering the application of sequential and parallel processing in the SP system, we need to distinguish between the workings of the SP machine and real-world processes (such as baking a cake or managing an industrial plant) to which the SP system may be applied:

- *The workings of the SP machine*. In the projected SP machine [2] (Section 3.2), it is envisaged that, while the ordering of symbols would be respected when one pattern is matched with another, the process of matching two patterns may be done left-to-right, right-to-left, or with many symbols matched in parallel. Likewise, many pairs of patterns may be matched in parallel.
- *Sequencing and parallelism in real-world processes*. Although the SP machine may be flexible in the way parallelism is applied to the matching of patterns, the system needs to respect sequencing and parallelism in the real-world processes to which the system may be applied. In broad terms, this may be achieved via the SP patterns themselves:
    - A sequence of operations may be modeled via the left-to-right sequence of symbols in a one-dimensional SP pattern, with "subroutines" if required, as outlined in Section 6.6.1.

- Since SP patterns at the "top" level are independent of each other, they may serve to model processes that may run in parallel.
- Naturally, a realistic model of any real-world project is likely to need additional information about such things as timings, dependencies between processes, and resources required.

6.6.4. Automatic Programming

With some improvements to the learning processes in the SP model (Section 9.5 in [1]; Section 5.1.4 in [2]), there is potential to by-pass at least some of the labor-intensive and costly hand-crafting of computer programs which is currently the mainstay of software engineering:

- *Learning of ontologies*. Knowledge of significant categories—people, buildings, vehicles, and so on—may be built up incrementally from information that users supply to the system. There is potential for the automatic structuring of such information in terms of class hierarchies and part-whole hierarchies.

- *Learning of procedures*. In a similar way, procedural knowledge may be built up from information supplied by users, and, via learning, such procedural knowledge may be structured automatically and generalized [28]. Likewise, for "programming by demonstration" in robotics [29].

Since knowledge in the SP system is intended to be transparent and comprehensible by people [2] (Section 2.4), it should be relatively straightforward to make adjustments and corrections to learned structures.

6.6.5. Verification and Validation

As was suggested in Section 5, the SP system may reduce or eliminate MUP instructions and their repetition in different applications. To the extent that that proves possible, we may reduce the opportunities for errors to be introduced. There is a corresponding potential to reduce or eliminate the need for "verification" in software development, and for improvements in the quality of software.

The SP machine also has potential in "validation": helping to ensure that what is being developed is what is required. If domain-specific knowledge, including the requirements for a software system, can be loaded directly into an SP machine, without the need for the traditional kind of programming, this will help to ensure that what the system does is what the users want. Users may have tighter control than is traditional over what a system does, with more freedom to make changes when required [30].

6.6.6. Technical Debt

By reducing the complexity of the software development process and by enabling users to have more direct control over how things work, the SP machine may also help to reduce or eliminate the problem of "technical debt", the way in which conventional systems can accumulate a backlog of "housekeeping" tasks and become unmanageably complex with the passage of time.

6.7. Information Compression

Given the central importance of information compression in the SP theory, one would naturally expect the SP machine to prove useful in that area of application. In effect, the multiple alignment framework serves, *inter alia*, to integrate three techniques for information compression: *chunking-with-codes*, *schema-plus-correction*, and *run-length coding* [1] (Section 2.2.9).

There is potential for the SP system to achieve relatively high levels of lossless compression for two main reasons:

- It is intended that, normally, the SP machine will perform a relatively thorough search of the space of alternative unifications of patterns and achieve correspondingly high levels of compression.

- If, as anticipated with some further development, the system will be able to learn discontinuous dependencies in data (see [2] (Section 3.3)), it will tap into sources of redundancy that appear to be outside the scope of traditional methods for compression of information.

In general, information compression can bring benefits in: economies in storage; speeding up transmission of information or reducing the demand for bandwidth, or some combination of the two (Section 6.7.1); providing a basis for inference (as described in [2] (Sections 2.1, 4.4, 10), and elsewhere in that article); and greater efficiency in processing with corresponding gains in energy efficiency (Section 6.7.2).

6.7.1. Economies in the Transmission of Information

A compressed version of a body of information, *I*, may be seen to comprise two parts:

- A "grammar" for *I* (which we may call *G*) containing patterns, at one or more levels of abstraction, that occur repeatedly in *I*.

- An "encoding" of *I* in terms of the grammar (which we may call *E*), including non-repeating information in *I*.

Where there is significant redundancy in *I*—which is true of most kinds of natural-language text and most kinds of images, especially moving images—*E* is likely to be much smaller than *G*. A grammar for *I* may also support the lossless compression of other bodies of information, provided they contain the same kinds of structures as *I*.

These things can provide the means of transmitting information very economically:

- A receiver (such as a TV set or a computer receiving web pages over the Internet) may be equipped with a grammar for the kind of information it is designed to receive, and some version of the SP system.

- Instead of transmitting "raw" data, or data that has been compressed in the traditional manner (containing both *G* and*E*), the encoding by itself would be sufficient.

- Using the SP system, the original data may be reconstructed fully, without any loss of information, by decoding the transmitted information (*E*) in terms of the stored copy of *G*.

A simple version of this idea, using a dictionary of words as a "grammar" for ordinary text, is already recognized (e.g., [31] and Chapter 11 in [32]). The apparent advantage of the SP system is the creation of more sophisticated grammars yielding higher levels of compression, and the application of the idea to kinds of information—such as moving images—where an ordinary dictionary would not work.

Since ordinary TVs work well without this kind of mechanism, some people may argue that there is no need for anything different. But:

- The growing popularity of video, TV and films on mobile services is putting pressure on mobile bandwidth [33], and increases in resolution will mean increasing demands for bandwidth via all transmission routes.
- It is likely that some of the bandwidth for terrestrial TV will be transferred to mobile services [34], creating an incentive to use the remaining bandwidth efficiently.
- There would be benefits in, for example, the transmission of information from a robot on Mars, or any other situation where a relatively large amount of information needs to be transmitted as quickly as possible over a relatively low-bandwidth channel.

6.7.2. Potential for Gains in Efficiency in Processing and in the Use of Energy

If, for example, we wish to search a large body of data, *I*, for instances of a pattern like "Treaty on the Functioning of the European Union", and to conduct similar searches for other patterns, there can be savings in processing if:

1) *I* is compressed by finding recurrent patterns in *I*, like "Treaty on the Functioning of the European Union", and replacing each instance of a given pattern by a relatively short code such as "TFEU", or, more generally, compressing *I*via grammar discovery and encoding as described in [1] (Chapter 9 and Section 3.5) and [2] (Sections 4 and 5).
2) Searching is done with "TFEU" instead of "Treaty on the Functioning of the European Union", or, more generally, searching is done with a compressed representation of the search pattern via the encoding processes in the SP system.

Although the first "encoding" step requires processing, that one-time cost can be more than offset by consequent savings in many searches. The savings arise because of the reduced size of *I* and because of the relatively small size of a code like "TFEU" compared with the pattern that it represents.

Since searching lies at the heart of how the SP system works, gains in the efficiency of searching should mean gains in the overall efficiency of processing in the SP system. There should be corresponding reductions in the use of energy.

6.8. Medical Diagnosis

The way in which the SP system may be applied in medical diagnosis is described in [35]. The expected benefits of the SP system in that area of application include:

- A format for representing diseases that is simple and intuitive.
- An ability to cope with errors and uncertainties in diagnostic information.
- The simplicity of storing statistical information as frequencies of occurrence of diseases.
- The system provides a method for evaluating alternative diagnostic hypotheses that yields true probabilities.
- It is a framework that should facilitate the unsupervised learning of medical knowledge and the integration of medical diagnosis with other AI applications.

The main emphasis in [35] is on medical diagnosis as pattern recognition. But the SP system may also be applied to causal diagnosis ([1] (Section 7.9) and [2] (Section 10.5)) which, in a medical context, may enable the system to reason that "The patient's fatigue may be caused by anemia which may be caused by a shortage of iron in the diet".

6.9. Managing "Big Data" and Gaining Value from It

The SP system has potential in the management and analysis of the large volumes of data that are now produced in many fields ("big data"), as discussed in [36] and summarized here.

The system may help overcoming the problem of variety in big data: the many different formats for knowledge and consequent problems in analysis. In view of the versatility of SP patterns, within the SP framework, to represent diverse kinds of knowledge [2] (Section 7), there is potential for them to serve as a universal format for knowledge and to facilitate the processing of diverse kinds of knowledge. The SP framework may also provide a mechanism for translating the original data into the universal format.

In the interpretation of data, the SP system has capabilities that include such things as the parsing and production of language, pattern recognition, and various kinds of probabilistic reasoning. But potentially the most useful facility with big data would be scanning for patterns, with recognition of family-resemblance or polythetic categories, at multiple levels of abstraction and with part-whole hierarchies, with inductive prediction and the calculation of associated probabilities, with a role for context in recognition, and robust in the face of errors of omission, commission or substitution. There are potential applications in several areas including, security, finance, meteorology, and astronomy.

With solutions to residual problems in unsupervised learning [2] (Sections 3.3 and 5.1.4), it seems likely that the SP machine, via the "DONSVIC" principle [2] (Section 5.2), will prove useful in discovering "interesting" or "useful" structures in big data, including significant entities, class-inclusion structures, part-whole structures, rules, regularities and associations, and discontinuous dependencies in data. The use of one simple format for different kinds of knowledge is likely to facilitate the learning of concepts that combine diverse kinds of knowledge.

Because of its potential for information compression (Section 6.7), the SP system may achieve useful reductions in the volume of big data and thus facilitate its storage and management. There is potential for substantial economies in the transmission of big data (Section 6.7.1), and for the efficient processing of compressed data, without the need for decompression (Section 6.7.2). And unlike conventional systems for compression of information, the SP system has capabilities in several aspects of intelligence.

6.10. Other Areas of Application

Some other areas of application where there is potential for the SP system are grouped here because, so far, that potential has not yet been much explored. But any of them may turn out to be significant.

6.10.1. Knowledge, Reasoning, and the Semantic Web

The SP framework may contribute to the development of the "semantic web"—a "web of data" to provide machine-understandable meaning for web pages [37,38,39]. In this connection, the main attractions of the system appear to be:

- *Simplicity, versatility and integration in the representation of knowledge*. The SP system combines simplicity in the underlying format for knowledge with the versatility to represent several different kinds of knowledge—and it facilitates the seamless integration of those different kinds of knowledge [2] (Section 7).

- *Versatility in reasoning*. The system provides for several kinds of reasoning and their inter-working (Chapter 7 in [1] and Section 10 in [2]).

- *Natural language understanding*. As suggested in Section 6.2, the SP system has potential for the understanding of natural language. If that proves possible, semantic structures may be derived from textual information in web pages, without the need, in those web pages, for the separate provision of ontologies or the like.

- *Automatic learning of ontologies*. As suggested in [2] (Section 5.2), the SP system has potential, via the DONSVIC principle, for the extraction of interesting or useful structures from data. These may include the kinds of ontologies which have been a focus of interest in the development of the semantic web.

- *Uncertainty and vagueness*. The SP system is inherently probabilistic and it is robust in the face of incomplete information and errors of commission or substitution. These capabilities appear promising as a means of coping with uncertainty and vagueness in the semantic web [40].

6.10.2. Bioinformatics

As described in [2] (Section 4), a central idea in the SP system is the concept of *multiple alignment*, borrowed from bioinformatics but with important differences from how that concept is normally understood in research in biochemistry.

Notwithstanding those differences, or perhaps because of them, the SP system has potential for the analysis of DNA sequences and amino-acid sequences in any or all of the following ways:

- The formation of alignments amongst two or more sequences, much as in multiple alignment as currently understood in bioinformatics.
- The discovery or recognition of recurrent patterns in DNA or amino-acid sequences, including discontinuous patterns.
- The discovery of correlations between genes, or combinations of genes, and any or all of: diseases; signs or symptoms of diseases; or any other feature or combination of features of people, other animals, or plants.

6.10.3. Maintaining Multiple Versions and Parts of a Document or Web Page

The multiple alignment concept provides a means of maintaining multiple versions of any document or web page, such as, for example, versions in different languages. Parts which are shared amongst different versions—such as pictures or diagrams—may be kept separate from parts that vary amongst different versions—such as text in a specific language. As with other kinds of data, the framework provides the means of maintaining hierarchies of classes (with cross classification if that is required) and of maintaining part-whole hierarchies and their integration with class hierarchies [2] (Section 9.1).

Updates to high-level classes appear directly in all lower-level classes without the need for information to be copied amongst different versions.

Of course, these kinds of things can be done in other ways. The strength of the SP framework is its versatility, reducing the need for *ad hoc* solutions in different areas.

6.10.4. Detection of Computer Viruses

The detection of already-known computer viruses and other malware can be more subtle and difficult than simply looking for exact matches for the "signatures" of viruses. The offending code may be rather similar to perfectly legitimate code and it may be contained within a compressed ("packed") executable file.

Here, the SP system and its capabilities for pattern recognition and learning may prove useful:

- Recognition in the SP system is probabilistic, it does not depend on the presence or absence of any particular feature or combination of features, and it can cope with errors.
- By compressing known viruses into a set of SP patterns, the system can reduce the amount of information needed to specify viruses and, as a consequence, it can reduce the amount of processing needed for the detection of viruses.

6.10.5. Data Fusion

There are many situations where it can be useful to merge or integrate two or more sources of information, normally when there is some commonality amongst the different sources. For example, in ecological research there may be a need to integrate different sources of information about the movement around the world of whales or other creatures; and in the management of a website, there may be a need to consolidate a set of web pages that are saying similar things.

At first sight, the SP system appears to be tailor made for this area of application. After all, the merging of fully or partially matching patterns lies at the heart of the SP system. But, as with big data (Section 6.9), the variety of different formats for knowledge can pose problems.

A possible solution, as with big data, is to translate different kinds of knowledge into a universal format. As before, the SP framework may provide that universal format and a mechanism for the translation.

6.10.6. New Kinds of Computer

In research to develop new kinds of computer, such as optical or chemical computers, it is commonly assumed that the first step must be to create the optical or chemical equivalent of a transistor or logic gate, probably because of the significance of those kinds of structures in the evolution of electronic computers as we know them today.

But the SP theory, with its emphasis on information compression via the matching and unification of patterns, suggests new approaches to the development of new technologies for computing, perhaps by-passing some of the structures and concepts that have been prominent in the development of electronic computers.

A potential benefit is that, with that primary focus on the matching and unification of patterns, it may prove easier to take advantage of the potential for high levels of parallelism in optical or chemical computers, or new kinds of electronic computers.

6.10.7. Development of Scientific Theories

Although Occam's Razor is widely recognized as a touchstone of good science and, as noted earlier, John Barrow has written that "Science is, at root, just the search for compression in the world." [5] (p. 247), there seems to have been relatively little interest—apart from some work on principles of "minimum length encoding" (see, for example, [41,42,43])—in making a connection between the practice of information compression and the development of scientific theories.

A possible reason is that most compression algorithms yield an encoding of the target data which is, from a human perspective, remarkably opaque: the output is normally smaller that the input—a useful result—but the encoding does not normally contain structures or relations that are meaningful for people. As noted in [2] (Section 13), this is probably because compression algorithms are normally designed to be quick-and-dirty, aiming for speed on low-powered computers, with a low priority for finding structures that yield high levels of compression; and, in accordance with the previously-mentioned DONSVIC principle, it appears to be those structures that are most meaningful for people.

Perhaps this does not matter. In a "practicalities-not-insight" approach to artificial intelligence that appears to be favored by some researchers, results that are useful in practice are given a higher priority than "explanation" or "understanding" [44].

With the SP system, there appears to be potential for something different and ultimately more acceptable as a way doing science: the possibility of creating a "discovery machine" [45] (p. 108*ff.*), deriving scientific theories from data, automatically or semi-automatically, and for obtaining analyses that people will regard as meaningful and insightful and which are also useful and practical. There is also potential for the quantitative evaluation of theories, as mentioned in Section 4.

In this area, reasons for optimism include:

- An "up market" emphasis in the SP program on finding structures that are good in terms of information compression.

- The DONSVIC principle, mentioned above—the discovery of natural structures via information compression [2] (Section 5.2). Structures that are good in terms of information compression seem generally to be structures that people regard as natural and meaningful.

- The versatility of the multiple alignment concept in the representation and processing of structures that are meaningful to people.

7. SEAMLESS INTEGRATION OF STRUCTURES AND FUNCTIONS WITHIN AND BETWEEN DIFFERENT AREAS OF APPLICATION

This section emphasizes again the importance of how the SP system may integrate structures and functions both within and between different areas of application. Instead of what can sometimes be awkward marriages between incompatible systems, there can be smooth interworking of, for example, pattern recognition, reasoning, and learning; of different kinds of reasoning; of different kinds of knowledge; and so on.

As previously noted, this facilitation of integration arises from: the use of one simple format for different kinds of knowledge; the use of one system—the multiple alignment framework—for the processing of knowledge; and the over-arching principle of information compression by the matching and unification of patterns.

As we have seen, the system provides for the integration of syntax and semantics in natural language processing; for the same kind of processing to achieve both the analysis and the production of language; for the integration of class-inclusion hierarchies, part-whole hierarchies, and other forms of knowledge; and for the integration of different aspects of intelligence in any combination: natural language processing, pattern recognition, different kinds of reasoning, unsupervised learning, and so on.

These kinds of integration are probably necessary to achieve human-like versatility and flexibility in the workings of computers.

8. CONCLUSIONS

The SP theory of intelligence combines conceptual simplicity with descriptive and explanatory power in several areas, including concepts of "computing", the representation of knowledge, natural language processing, pattern recognition, several kinds of reasoning, the storage and retrieval of information, planning and problem solving, unsupervised learning, information compression, neuroscience, and human perception and cognition.

In the SP machine there is potential for the simplification of computing systems, including software, with corresponding savings in the time, effort and cost in the development of applications, and other benefits related to information compression.

As a theory with a broad base of support, the SP theory promises deeper insights and better solutions in several areas of application, including unsupervised learning, natural language processing, autonomous robots, computer vision, intelligent databases, software engineering, information compression, medical diagnosis, and big data; and perhaps also in areas such as the semantic web, bioinformatics, structuring of documents, the detection of computer viruses, data fusion, new kinds of computer, and the development of scientific theories.

The SP theory should facilitate the integration of structures and functions, both within a given area and amongst different areas—a likely pre-requisite for the achievement of human-like versatility and flexibility in the way computers work.

These potential benefits and applications are not merely of theoretical interest. If, as a conservative estimate, they were to add 5% to the value of annual worldwide IT investments [46], they would be worth $190 billion each year, and increasing with the continuing growth of IT.

As suggested in [2] (Section 3.2) and in Section 3 of this article, further development would be assisted by the creation of a high-parallel, open source version of the SP machine, available to researchers throughout the world to examine what can be done with the system and to create new versions of it.

ACKNOWLEDGMENTS

I am grateful to anonymous referees for many useful comments, and to Daniel Wolff for pointing out the potential of the SP system in the detection of computer viruses and in the management and analysis of big data.

CONFLICTS OF INTEREST

The author declares no conflict of interest.

REFERENCES

1. Wolff, J.G. *Unifying Computing and Cognition: The SP Theory and Its Applications*; CognitionResearch.org: Menai Bridge, UK, 2006.

2. Wolff, J.G. The SP theory of intelligence: An overview. *Information* **2013**, *4*, 283–341.

3. According to this criterion, a good theory in a given area lies somewhere between things which are simple but very general—Such as the laws of arithmetic—And over-specific "theories" that simply repeat what it is they are meant to explain.

4. Hence the name "SP".

5. Barrow, J.D. *Pi in the Sky*; Penguin Books: Harmondsworth, UK, 1992.

6. Similar things were said by Allen Newell about research in human perception and cognition, in his well-known essay on why "You can't play 20 questions with nature and win" ([47]. See also [48,49]). And writing about machine technologies to augment our senses, John Kelly and Steve Hamm say "... there's a strong tendence [for scientists] to view each sensory field in isolation as specialists focus only on a single sensory capability. Experts in each sense don't read journals devoted to the others, and they don't attend one another's conferences. ..." [45] (p. 74).

7. Turing, A.M. On computable numbers, with an application to the Entscheidungsproblem. *Proc. Lond. Math. Soc.* **1936**,*42*, 230–265.

8. Turing, A.M. On computable numbers, with an application to the Entscheidungsproblem. A correction. *Proc. Lond. Math. Soc.* **1936**, *42*, 544–546.

9. Turing, A.M. Computing machinery and intelligence. *Mind* **1950**, *59*, 433–460.

10. Dreyfus, H.L. *What Computers Can't Do: A Critique of Artificial Reason*; Harper and Row: New York, NY, USA, 1972; Revised in 1979.

11. Dreyfus, H.L. *What Computers Still Can't Do*; MIT Press: New York, NY, USA, 1992.

12. The latest version is SP70.

13. Even in systems such as DBMSs or expert-system shells, mentioned above, the domain-specific knowledge which is loaded into the system may be regarded as part of the software.

14. The reason for putting "CPU" in quote marks in the lower part of Figure 3 is that an SP machine may not have a discrete central processing unit like a conventional computer. The SP system is probably better suited to a "data-centric" style [45] (pp. 87–93), with a closer integration of data and processing than in conventional computers.

15. Wolff, J.G. Towards an intelligent database system founded on the SP theory of computing and cognition. *Data Knowl. Eng.* **2007**, *60*, 596–624.

16. Wolff, J.G. Learning syntax and meanings through optimization and distributional analysis. In *Categories and Processes in Language Acquisition*; Levy, Y., Schlesinger, I.M., Braine, M.D.S., Eds.; Lawrence Erlbaum: Hillsdale, NJ, USA, 1988; pp. 179–215.

17. Available online: www.cognitionresearch.org/lang learn.html (accessed on 19 December 2013).

18. McCawley, J.D. The Role of Semantics in a Grammar. In *Universals in Linguistic Theory*; Lecture Notes in Artificial Intelligence; Bach, E., Harms, R.T., Eds.; Holt, Rinehart and Winston, Inc.: New York, NY, USA, 1968; pp. 125–169.

19. See, for example, "Interlingual machine translation", Wikipedia, available online: http://bit.ly/RITcHU. retrieved 31 May 2013.

20. That number may be reduced if one or more languages are used as stepping stones to the translation of others.

21. This reference to autonomous robots does not in any way endorse or defend the unethical or illegal use of airborne drones or other autonomous robots to cause death, injury, or damage to property.

22. Archibald, C.; Altman, A.; Greenspan, M.; Shoham, Y. Computational pool: A new challenge for game theory pragmatics. *AI Mag.* **2010**, *31*, 33–41.

23. Wolff, J.G. Application of the SP theory of intelligence to the understanding of natural vision and the development of computer vision 2013. CognitionResearch.org, Menai Bridge, UK, unpublished work. 2013.

24. Szeliski, R. *Computer Vision: Algorithms and Applications*; Springer: London, UK, 2011.

25. As with information compression, a focus on the isolation of discrete objects in binocular vision is distinct from the more usual interest in the way that slight differences between the two images enables us to see the scene in depth.

26. Marr, D.; Poggio, T. A computational theory of human stereo vision. *Proc. R. Soc. Lond. B* **1979**, *204*, 301–328.

27. The Ames room illusion shows that human vision does not always yield an accurate knowledge of 3D structures (see, for example, Wikipedia, available online: http://en.wikipedia.org/wiki/Ames_room. retrieved 4 December 2013).

28. Related work includes "process mining" as in, for example, the "ProM" system, available online: http://bit.ly/13iS2Ch.

29. See, for example, "Programming by demonstration", Wikipedia, available online: http://bit.ly/121GrKJ. retrieved 6 June 2013.

30. Notice that the distinction, just referenced, between domain-specific knowledge and MUP instructions is different from the often-made distinction between declarative programming and imperative or procedural programming. It is envisaged that the SP machine would reduce or eliminate the MUP instructions in software but that it would be able to model both the declarative "what" and (Section 6.6.1) the procedural "how" of the real world.

31. Giltner, M.; Mueller, J.C.; Fiest, R.R. Data Compression, Encryption, and in-Line Transmission System. U.S. Patent 4,386,416, 31 May 1983.

32. Storer, J.A. *Data Compression: Methods and Theory*; Computer Science Press: Rockville, MD, USA, 1988.

33. See, for example, "Data jam threat to UK mobile networks", BBC News, 16 November 2012. available online: http://bbc.in/T5g14s.

34. See, for example, "TVs will need retuning again to make room for mobile services", The Guardian, 16 November 2012. available online: http://bit.ly/ZZ9N8C.

35. Wolff, J.G. Medical diagnosis as pattern recognition in a framework of information compression by multiple alignment, unification and search. *Decis. Support Syst.* **2006**, *42*, 608–625.

36. Wolff, J.G. The management and analysis of big data, and the SP theory of intelligence 2013. CognitionResearch.org, Menai Bridge, UK, unpublished work. 2013.

37. Berners-Lee, T.; Hendler, J.; Lassila, O. The Semantic Web. *Scientific American* **2001**, 35–43.

38. Shadbolt, N.; Hall, W.; Berners-Lee, T. The semantic web revisited. *IEEE Intell. Systems* **2006**, *21*, 96–101.

39. See also W3C information about the semantic web at http://bit.ly/1bX7Nv.

40. Lukasiewicz, T.; Straccia, U. Managing uncertainty and vagueness in description logics for the Semantic Web. *Web Semant. Sci. Serv. Agents World Wide Web* **2008**, *6*, 291–308.

41. Solomonoff, R.J. A formal theory of inductive inference. Parts I and II. *Inf. Control* **1964**, *7*, 1–22, 224–254.

42. Wallace, C.S.; Boulton, D.M. An information measure for classification. *Comput. J.* **1968**, *11*, 185–195.

43. Rissanen, J. Modelling by the shortest data description. *Autom. J. IFAC* **1978**, *14*, 465–471.

44. See, for example, "Not like us: Artificial minds we can't understand", New Scientist, 10 August 2013. available online:http://bit.ly/15m1cnm. As noted in Section 6.5 of this article, transparency in the SP system, in both the representation of knowledge and in reasoning, may prove useful where evidence needs to be examined in a court of law.

45. Kelly, J.E.; Hamm, S. *Smart Machines: IBM's Watson and the Era of Cognitive Computing*; Columbia University Press: New York, NY, USA, 2013.

46. Expected to be $3.8 trillion by 2014 ("Gartner: Big data will help drive IT spending to $3.8 trillion in 2014", InfoWorld, 3 January 2013. available online: http://bit.ly/Z00SBr).

47. Newell, A. You Can't Play 20 Questions with Nature and Win: Projective Comments on the Papers in This Symposium. In *Visual Information Processing*; Chase, W.G., Ed.; Academic Press: New York, NY, USA, 1973; pp. 283–308.

48. Newell, A. Précis of *Unified Theories of Cognition*. *Behav. Brain Sci.* **1992**, *15*, 425–437.

49. Newell, A. *Unified Theories of Cognition*; Harvard University Press: Cambridge, MA, USA, 1990.

Chapter 5

Artificial Intelligence for Speech Recognition Based on Neural Networks

Takialddin Al Smadi[1], Huthaifa A. Al Issa[2], Esam Trad[3], Khalid A. Al Smadi[4]

[1]Department of Communications and Electronics Engineering, College of Engineering, Jerash University, Jerash, Jordan

[2]Department of Electrical and Electronics Engineering, Faculty of Engineering, Al-Balqa Applied University, Al-Huson College University, Al-Huson, Jordan

[3]Departments of Communications and Computer Engineering, Jadara University, Irbid, Jordan

[4]Jordanian Sudanese Colleges for Science & Technology, Khartoum, Sudan

ABSTRACT

Speech recognition or speech to text includes capturing and digitizing the sound waves, transformation of basic linguistic units or phonemes, constructing words from phonemes and contextually analyzing the words to ensure the correct spelling of words that sounds the same. Approach: Studying the possibility of designing a software system using one of the techniques of artificial intelligence applications neuron networks where this system is able to distinguish the sound signals and neural networks of irregular users. Fixed weights are trained on those forms first and then the system gives the output match for each of these formats and high speed.

The proposed neural network study is based on solutions of speech recognition tasks, detecting signals using angular modulation and detection of modulated techniques.

KEYWORDS

Speech Recognition, Neural Networks, Artificial Networks, Signals Processing

1. INTRODUCTION

Artificial intelligence applications have proliferated in recent years, especially in the applications of neural networks where they represent an appropriate tool to solve many problems highlighted by distinguished styles and classification.

The year of 1943 is known as the beginning of the evolution of artificial neural systems.

The first formal model of neurons through a computer model that includes all the necessary elements and the completion and implementation of the electronic form of this model is not practical or reasonable in terms of tech during the vacuum tube. It should be noted that this model has been applied extensively to describe computer hardware for the vacuum tube [1] . Initially, planned tutorial to update connections of nerve cells that are referred to the law educational learning rule HYIP has stated that the information can be stored in the links and connections. It is recognized that learning technology has proved its benefits in the future development of this field. Hip education Act initial contribution in neural network theory had been built and tested in the first study of the neurological computer in the 1950s, where the application contacts automatically and during this stage the term preceptor called the unit represented for neural cell to invent the term world and divorced on the neuron, he pioneered the term frank Rosenblatt in 1958. This invention was a viable training machine learning and classification of certain models by modulating communication components first. In this way it has become along with the imagination of engineers and scientists and a background to the calculations of this type of machinery which is still used today.

In the early 1960s, a new created method called Adaptive Linear Combiner developed a very useful law [2].

2. PATTERN RECOGNITION

Automatic recognition, description, classification and grouping patterns are important parameters in various engineering and scientific disciplines such as biology, psychology, medicine, marketing, computer vision, artificial intelligence and remote sensing. The template can be fingerprint images, handwritten words cursive, a human face or the voice signal. Given the pattern, its recognition/classification may be one of the following two tasks [3] .

a) Under the supervision of a classification, discriminated analysis, in which the input pattern is defined as a member of a predefined class;

b) Unsupervised classification, clustering in which is the class template is unknown.

Recognition of the problem here is as a classification or classification problems, where the classes are defined by either the system designer in a controlled classification or learned based on similar models in unsupervised classification.

These applications include data mining the definition of “plan”. For example, he correlations or independently in millions of multidimensional models, document classification effectively search text documents, financial, forecasting, organization and retrieval of multimedia databases and biometrics. The rapidly growing and available computing power, enabling faster processing of huge amounts of data, also promoted the use of complex and diverse methods for classification and analysis of data. At the same time, the demand for automatic pattern recognition is growing due to the presence of large databases and strict requirements speed, accuracy and cost. Design of recognition system template essentially consists of the following three aspects:

a) Collection and preprocessing, data reporting;

b) Decision-making process;

c) Scope dictates the choice of pretreatment technique.

Schema view and decision making models It is recognized that the problem of clearly defined and sufficiently limited recognition will lead to the introduction of the compact model and simple decision-making strategy. Learning from a set of examples is an important and necessary attribute of most systems of recognition template.

The most prominent approaches for pattern recognition are:

a) Matching pattern;

b) Statistical classification;

c) Syntactic or structural conformity and neural networks.

3. NEURAL NETWORKS

Neural networks consist of a set of nodes that a special type of account collectively and that each node is the standard unit of account and the contract could work in parallel depends on the interactions among themselves and how they relate to some of the scholars are defined as:

Ÿ Mathematical models simulating characteristics of biological systems that deal with information in parallel composed of relatively simple elements called.

Ÿ Is a simple entity class of algorithms that are formulated in charts (graphs grouped these schemes a large number of algorithms and these algorithms provide solutions to a number of complex problems [4] .

To highlight the activity of neural networks is the process of classification and coding and to highlight the properties of neural networks are:

a) Resistance to noise;

b) Flexibility in dealing with the distorted images;

c) Maximum resistance to tag images of dismembered or partially decomposed;

d) Combinations of parallel processes with a large number of operating units that stimulate by interdependence of processes in addition to the stock of information distributed in parallel.

With non-linear operations, i.e. their ability to make non-linear relationships include maps of noise that makes them a good source of ratings and attribution (classification predication);

e) High capacity to adapt the system of logarithms and powers of education internal allows the use of internal adjustment that lives in the vicinity of lasting change.

Types of Neural Networks

Possible to identify the most common types of neural networks with input types and learn some common uses as in Table 1 shown [5] [6].

4. PROCEDURE WORKS

The method consists of iteratively selecting the most distant score with respect to mean. If this score goes beyond a certain threshold, the score is removed and mean and standard deviation estimations are recalculated. When there are only a few utterances to estimate mean and variance, this method leads to a great improvement. Text dependent and text independent experiments have been carried out by using a telephonic multisession database. The paper presents the inter-relationship between algorithmic research system developments based on the experience from the speaker using mini-problems during the system design process, and presents a model of speech recognition based on artificial neural networks [7] . Figure 1 shows the diagram of the processing of speech signals.

a) Present study of artificial neural networks for speech recognition task. Neural network size influence on the effectiveness of detection of phonemes in words. The research methods of speech signal parameterization. Learn about how to use linear prediction analysis, a temporary way of learning of the neural network for recognition of phonemes. The proposed way of teaching as input requires only the transcription of words from the training set and do not require any manual segmentation of words;

b) Development and research of the methods for diagnosing and detecting modulated signals;

c) Software implementation and pilot testing on real signals of neural network methods for processing.

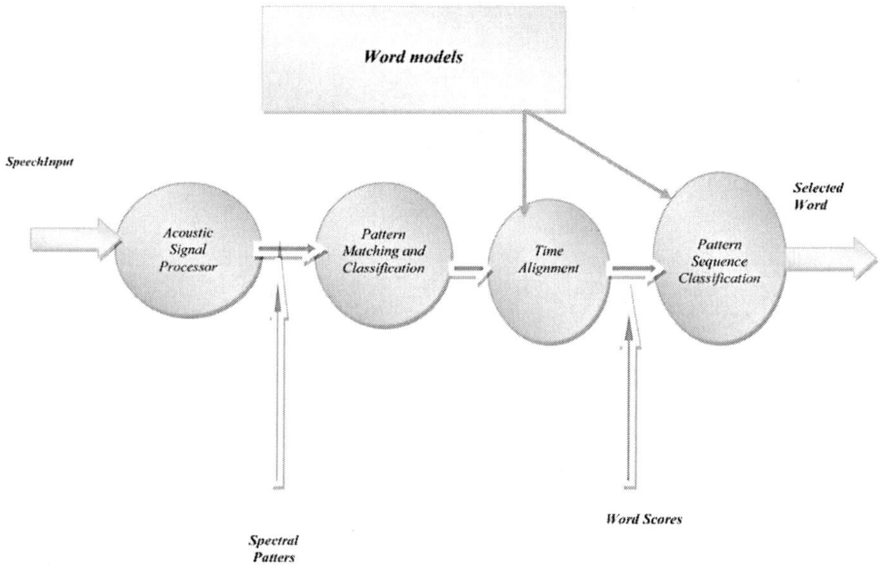

Figure 1. Diagram of the processing of speech signals planning.

4.1. Recognition Process Recognition Algorithm

Ÿ Input signal into the computer and select word boundaries;

Ÿ Allocation of parameters characterizing the signal spectrum;

Ÿ The use of artificial neural network to evaluate the degree of proximity of acoustic parameters;

Ÿ Comparison with standards in the dictionary [8] .

Table 1. Types of neural networks and application.

Common uses	Input method	Input type	Types of neural networks
Associated memory to distinguish ASCII characters	Supervised	Binary	**Hopfield-Net**
Connect with similar dual channel	Supervised	Binary	**Hammin_Net**
Assembly (adaptive resonance theory)	Supervised	Binary	**Carpenter/grassbery classifier**
Discrimination and classification of simple shapes	Supervised	Continuous	**Perceptron**
Featuring complex shapes and classification	Supervised	Continuous	**Multi-layer perceptron**
Evaluation of vector and speech, and analogy to biological neural networks	**Supervised**	**Continuous**	**Kohonenself organizing feature map**

Voice signal as an input to a neural network, after processing the audio data received an array of segments of the signal. Each segment corresponds to a set of numbers that characterize the amplitude spectra of a signal, to prepare for the calculation for the signal outputs of the neural network to write all the numbers shows in Table 2, where a row which is a set of numbers of each frame.

Where I is the number of values of a set of numbers, N is the number of sets of numbers (frame signal after slicing). The number of input and output neurons is known, each of the input neurons corresponds to one set of numbers, and the output layer only one neuron, which corresponds to the desired value of the signal recognition. Table 3 shows the parameter definition uses in this research as shown in Figure 2.

4.2. Equations

To calculate the output of the neural network, it's a must complete the following successive steps [9] :

Step 1: Initiate all contexts of all the neurons in the hidden layer;

Step 2: Apply the first set of numbers to the neural network. Calculate the output of the hidden layer.

$$y_j = f\left(\sum_{i=1}^{I} \omega_{ij} X_{1i} + \beta_i + \omega_j X_j\right) \tag{1}$$

F(x)—non-linear activation function

$$y_j = \frac{1}{1+\mathrm{e}^{-aS_j}} \tag{2}$$

for the numbers from 0 to 9.

To recognize the one number you need to build your own neural network it's a must to build 10 of neural networks. Database of over 250 words (numbers from 0 to 9) with different variations of pronunciation, base randomly divided into two equal parts-tutorial and sample tests. When training neural network recognition of one number, for number 5, the desired output of the neural network needs to be unit for the training set with the number 5 and the remainder is zero.

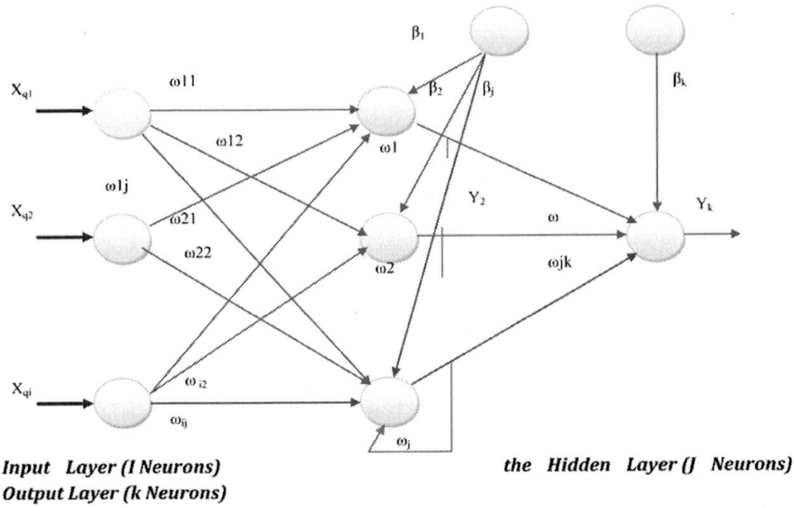

Figure 2. The structure of a neural network with a feedback.

Neural network training is carried out through the consistent presentation of the training set, with simultaneous tuning scales in accordance with a specific procedure, until around the variety of configuration error reaches an acceptable level. Error in the system function will be calculated by the following formula:

$$E = \frac{1}{2N} \sum_{i=1}^{N} (y_{ki} - d_i)^2 \tag{3}$$

where N is the number of training samples processed by neural network examples the real output of the neural network.

Table 2. Description of a set of speech signal.

Frame	1-value	2-value	…	I-value
1-Frame	x_{11}	x_{12}	…	x_{1I}
2-Frame	x_{21}	x_{22}	…	x_{2I}
…	…	…	…	…
N-Frame	x_{N1}	x_{N2}	…	x_{NI}

Table 3. Parameters definition.

Name	Definition
x_{qi}	*i*-th *q* is the input value to a set of numbers
y_j	Output *j*-neuron layer
ω_{ij}	The weight of the link connecting the *i*-th neuron with the *j*-th neuron
ω_j	weight feedback
β_j	Weight feedback *j*-th neuron; the offset of the *j*-the neuron layer

A prototype of a neuron is nerve cell biology. A neuron consists of a cell body, or soma, and two types of external wood-like branches: Axon and dendrites. The cell body contains the nucleus, which holds information on hereditary characteristics and plasma with molecular tools for the production and transmission of elements of the neuron of the necessary materials. A neuron receives signals from other neurons through the dendrites and transmits signals generated by the cells of the body, along the axon, which at the end of branches into the fiber, the endings of synapses [1] [3] .

Mathematical model of a neuron described democratic ratio:

$$y = f(s), swx_i w_i wb \tag{4}$$

where w_i is the synapse, the weight (b)-offset value, s is the input signal, y-signal output neuron, n is the number of inputs to the neuron, f-function is activated. Technical model of a neuron is represented in Figure 3.

Block diagram of a neuron: $x_1, x_2, \cdots, x_n$-input neuron; $w_1, w_2, \cdots,$ the W_n-a set of weights; F(S) is a function of activation; y-output signal, neuro control performs simple operations like weighted summation, treating the result of nonlinear threshold conversion. Feature of neural network approach is that the structure of the simple homogeneous elements allows you to meet the challenges of the complex relationships between items. The structure of relations defines the functional properties of the network as a whole.

The functional features of neurons and how they combine into a network structure determines the features of neural networks. To meet the challenges of the most adequate identification and management are multilayer neural networks direct action or layered perceptions. When designing neurons together in layers, each of which handles vector signals from the previous layer. Minimum implementation is smiling two-layer neural network, consisting of the input (switch gear), intermediate (hidden), and the output layer [10] (Figure 4).

Implementation of the model of two-layer neural network of direct action has the following mathematical representation:

$$y(\theta) = F_i\left(\sum_{j-1}^{nh} W_{ij} f_i\left(\sum_{j-1}^{nh} w_{ij}\phi_j + w_{jo}\right) + W_{jo}\right) \tag{5}$$

where the dimension of the vector inputs is: nϕ ϕ neural network; nh-the number of neurons in the hidden layer; θ-vector of the configurable parameters of the neural network, which includes weights and neuron-by offset (w_{ji}, W_{ij}); $f_j(x)$-activation function for the hidden layer neurons; $F_i(x)$-activation function neuron in the output layer.

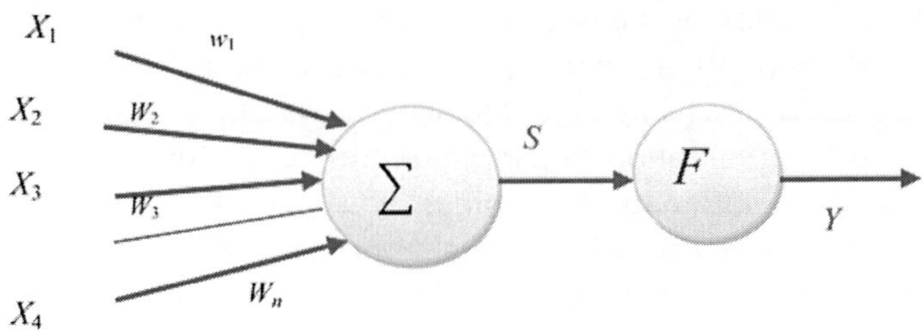

Figure 3. Technical model of a neuron is represented.

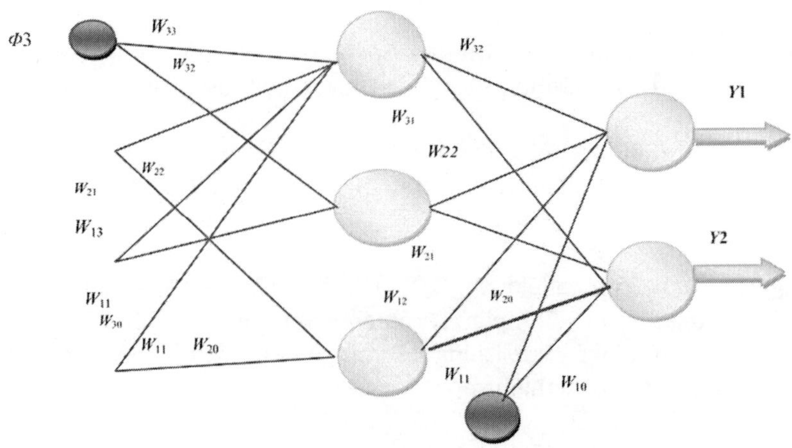

Figure 4. Structural diagram of two-layer neural network.

The most important feature of neural network method is the possibility of parallel processing. This feature if there are a large number of international neural connections enables to significantly accelerate the process of signet-data processing [6] . A possibility of processing of speech signals in real time. The neural network has qualities that are inherent in the so-called artificial intelligence [11].

5. CONCLUSION

Model of speech recognition was based on artificial neural networks. This was investigated to develop a learning neural network using genetic algorithm. This approach was implemented in the system identification numbers, coming to the realization of the system of recognition of voice commands. A system of automatic recognition of speech keywords that were associated with the processing of telephone calls or a sphere of security was developed. The accuracy level of forecasting on the basis of present data set experience was always better.

REFERENCES

1. Childer, D.G. (2004) The Matlab Speech Processing and Synthesis Toolbox. Photocopy Edition, Tsinghua University Press, Beijing, 45-51.

2. Chien, J.T. (2005) Predictive Hidden Markov Model Selection for Speech Recognition. IEEE Transaction on Speech and Audio Processing, 13.

3. Luger, G. and Stubblefield, W. (2004) Artificial Intelligence: Structures and Strategies for Complex Problem Solving. 5th Edition, The Benjamin/Cummings Publishing Company, Inc.http://www.cs.unm.edu/~luger/ai-final/tocfull.htm

4. Choudhary, A. and Kshirsagar, R. (2012) Process Speech Recognition System Using Artificial Intelligence Technique. International Journal of Soft Computing and Engineering (IJSCE), 2.

5. Ovchinnikov, P.E. (2005) Multilayer Perceptron Training without Word Segmentation for Phoneme Recognition. Optical Memory & Neural Networks (Information Optics), 14, 245-248.

6. Guo, X.Y., Liang, X. and Li, X. (2007) A Stock Pattern Recognition Algorithm Based on Neural Networks. Third International Conference on Natural Computation, 2.

7. Dai, W.J. and Wang, P. (2007) Application of Pattern Recognition and Artificial Neural Network to Load Forecasting in Electric Power System. Third International Conference on Natural Computation, 1.

8. Shahrin, A.N., Omar, N., Jumari, K.F. and Khalid, M. (2007) Face Detecting Using Artificial Neural Networks Approach. First Asia International Conference on Modelling & Simulation.

9. Lin, H., Hou, W.S., Zhen, X.L. and Peng, C.L. (2006) Recognition of ECG Patterns Using Artificial Neural Network. Sixth International Conference on Intelligent Systems Design and Applications, 2.

10. Al Smadi, T.A. (2013) Design and Implementation of Double Base Integer Encoder of Term Metrical to Direct Binary. Journal of Signal and Information Processing, 4, 370.

11. Takialddin Al Smadi Int. An Improved Real-Time Speech Signal in Case of Isolated Word Recognition. Journal of Engineering Research and Applications, 3, 1748-1754.

Chapter 6

Visual descriptors for scene categorization: experimental evaluation

Xue Wei1, Son Lam Phung1, Abdesselam Bouzerdoum1

School of Electrical, Computer, and Telecommunications Engineering, University of Wollongong, Northfields Av, Wollongong, NSW 2522, Australia

ABSTRACT

Humans are endowed with the ability to grasp the overall meaning or the gist of a complex visual scene at a glance. We need only a fraction of a second to decide if a scene is indoors, outdoors, on a busy street, or on a clear beach. In recent years, computational gist recognition or scene categorization has been actively pursued, given its numerous applications in image and video search, surveillance, and assistive navigation. Many visual descriptors have been developed to address the challenges in scene categorization, including the large number of semantic categories and the tremendous variations caused by imaging conditions. This paper provides a critical review of visual descriptors used for scene categorization, from both methodological and experimental perspectives. We present an empirical study conducted on four benchmark data sets assessing the classification accuracy and class separability of state-of-the-art visual descriptors.

KEYWORDS

Visual descriptors, Scene categorization, Survey and evaluation Gist recognition

1. INTRODUCTION

Humans can grasp rapidly the overall meaning of a complex visual scene. With a single glance, they can determine whether they are looking at a room, a beach, or a forest (Pavlopoulou and Yu 2010). Viewers need only a fraction of a second to associate a picture with an abstract concept such as girl clapping or busy street (Goldstein 2010). This ability to understand the conceptual meaning of a scene at a glance, regardless of its visual complexity and without attention to details, is known as *gist recognition*. The gist of a scene plays many roles in visual perception. It guides the viewer's attention, aids object recognition, and affects the viewer's recollection of the scene (Snowden et al. 2004).

Understanding gist recognition in humans and replicating this ability in computers has been an ongoing research goal (Fei-Fei et al. 2007b; Loschky et al. 2007; Wei et al. 2010; Walther et al. 2011; Lee and Choo2013; Linsley and MacEvoy 2014; Malcolm et al. 2014). In the early psychovisual studies, Potter found that a scene is processed rapidly at an abstract level before intentional selection occurs (Potter 1975). Gist recognition is shown to happen in the first 30–300 ms of viewing a scene (Goldstein 2010). This perception occurs much earlier than the time required for recognizing individual objects in the scene. Consequently, gist recognition is considered to rely significantly more on holistic and low-level properties than on the detection of individual objects (Loschky et al. 2007). These low-level properties include color (Castelhano and Henderson 2008; Oliva 2000), edges (Pavlopoulou and Yu 2010), and texture (Renninger and Malik 2003). It was even found that object shape and identity are not necessary for the rapid perception of scenes (Oliva and Torralba 2001).

In computer vision, gist recognition is also known as *scene categorization,* which aims to classify a scene into semantic categories (Lazebnik et al. 2006; Xiao et al. 2014a; Krapac et al. 2011; Yao et al. 2012; Zhao and Xing 2014). The scene could be a static image or dynamic video, and the semantic categories could be indoor versus outdoor, gas station versus restaurant, or slow traffic versus flowing traffic. Scene categorization can be used to provide cues about objects and actions, detect abnormal events in public places, sense dangerous situations, and search for images and video; therefore, it is highly useful for applications in surveillance (Zhang et al. 2008; Gowsikhaa et al. 2014), navigation (Chang et al. 2010; Jia et al. 2014; Khosla et al. 2014; Fuentes-Pacheco et al. 2015), and multimedia (Chella et al. 2001; Maggio and Cavallaro 2009; Huang et al. 2008; Zhou et al. 2010; Dai et al. 2015).

In computational scene categorization, visual descriptors play a central role in recognition performance. A good visual descriptor must possess discriminative power to characterize different semantic categories, while remaining stable in the presence of inter- and intra-class variations caused by photometric and geometric image distortions. Note that previously Douze et al. (2009) compared different visual descriptors for image search, but the compared descriptors were restricted to only GIST descriptor (Oliva and Torralba2001) and its variants. Mikolajczyk and Schmid (2005) evaluated the local descriptors, such as SIFT-based features, steerable filters, complex filters, and moment invariants. Van de Sande et al. (2010) analyzed mainly SIFT-based color descriptors on the PASCAL VOC Challenge 2007. Xiao et al. (2014a) compared fourteen descriptors on the SUN397 data set. Their comparison also focused on the local features, such as*scale-invariant feature transform* (SIFT) (Lowe 2004), *histogram of oriented gradients* (HOG) (Dalal and Triggs 2005), and *local binary pattern* (LBP) (Ojala et al. 1996).

This paper aims to assess the state-of-the-art visual descriptors for scene categorization of static images, from both methodological and experimental perspectives. The compared descriptors range from biologically-inspired features, local features to global features. The paper is structured as follows. Section 2 reviews visual descriptors used for scene categorization. Section 3 describes the public data sets and evaluation protocols for scene categorization. Section 4 presents results of experimental evaluations on four benchmark data sets. Section 5 gives concluding remarks.

2. VISUAL DESCRIPTORS

Many algorithms have been developed for scene categorization from static images. For convenience, the list of the most widely used acronyms in this paper is given in Table 1. In most algorithms, visual features are first extracted from the input image, and then classified into semantic categories using a trained classifier, e.g., support vector machines. Hence, feature extraction plays a vital role in scene categorization. The existing approaches for visual feature extraction in scene categorization can be divided into three broad categories: biologically-inspired methods, local features, and global features. Table 2 summarizes the three categories and gives the representative visual descriptors of each category.

Table 1. List of acronyms used in this paper

Acronym	Definition
BIF	Biologically-inspired features (Song and Tao 2010)
BOP	Bag-of-parts (Juneja et al. 2013)
BoW	Bag-of-words (Fei-Fei et al. 2007a)
CENTRIST	Census transform histogram (Wu and Rehg 2011)
CNN	Convolutional neural network (LeCun et al. 1998; Krizhevsky et al. 2012)
DeCAF	Deep convolutional activation feature (Donahue et al. 2014)
GIST	An abstract representation of the scene (Oliva and Torralba 2001)
HIK	Histogram intersection kernel (Maji et al. 2008)
HMAX	Hierarchical model and X (Riesenhuber and Poggio 1999)
HOG	Histogram of oriented gradients (Dalal and Triggs 2005)
HOG-SPM	HOG with spatial pyramid matching (Lazebnik et al. 2006)
HSOG	Histogram of the second-order gradients (Huang et al. 2014a)
LBP	Local binary pattern (Ojala et al. 1996)
LBP-HF	LBP with Fourier histogram (Ahonen et al. 2009)
LCS	Local color statistic (Clinchant et al. 2007)
OB	Object bank (Li et al. 2010)
PLBP	LBP with pyramid representation (Qian et al. 2011)
RBF	Radial basis function kernel (Scholkopf et al. 1997)
SEV	Sequential edge vectors (Morikawa and Shibata 2012)
SIFT	Scale-invariant feature transform (Lowe 2004)
SIFT-FV	SIFT with Fisher vector (Sanchez et al. 2013)
SIFT-LLC	SIFT with locality-constrained linear coding (Wang et al. 2010)
SIFT-ScSPM	SIFT with sparse coding based spatial pyramid matching (Yang et al. 2009)
SIFT-SPM	SIFT with spatial pyramid matching (Lazebnik et al. 2006)
SURF	Speeded up robust features (Bay et al. 2008)
SVM	Support vector machine (Burges 1998)

Table 2. Classification of scene categorization descriptors

Approaches	Representative works
Biologically-inspired feature extraction	
Visual cell model	HMAX (Serre et al. 2005)
Layout properties	GIST (Oliva and Torralba 2001)
Deep learning features	Convolutional neural networks (Zhou et al. 2014)
	OverFeat (Sermanet et al. 2014)
	DeCAF Donahue et al. (2014)
Local feature extraction	
Patch-based features	SIFT (Lowe 2004)
	HOG (Dalal and Triggs 2005)
	LBP (Ojala et al. 1996)
	CENTRIST (Wu and Rehg 2011)
	HSOG (Huang et al. 2014a)
Object-based model	Object bank (Li et al. 2010)
Region-based model	Edge vectors (Morikawa and Shibata 2012)
	Bag of parts (Juneja et al. 2013)
Keypoint-based features	SURF (Bay et al. 2008)
	FREAK (Alahi et al. 2012)
	BRISK (Leutenegger et al. 2011)
Global feature formation	
Principal component analysis	SIFT features with PCA (Ke and Sukthankar 2004)
Histogram	Multi-resolution histogram (Hadjidemetriou et al. 2004)
Feature encoding	SPM (Lazebnik et al. 2006)
	ScSPM (Yang et al. 2009)
	LLC (Wang et al. 2010)
	FV (Sanchez et al. 2013; Dixit et al. 2015)

2.1 Biologically-inspired feature extraction models

To mimic the gist recognition capability of the human vision, researchers have developed a variety of computational algorithms for scene categorization from static images. For example, Lee and Mumford (2003) developed a hierarchical Bayesian inference for scene reconstruction based on the *early visual neurons*. Song and Tao (2010) suggested a gist recognition model, where intensity, color, and *C*1 visual features are extracted. Grossberg and Huang (2009) proposed the ARTSCENE system based on gist and texture features for natural scene classification. In the following subsections, we discuss three major biologically-inspired feature extraction approaches: the HMAX model, the GIST model, and the deep learning model.

2.1.1 HMAX model

Riesenhuber and Poggio (1999) proposed a feed-forward architecture, called HMAX, which is inspired by the hierarchical nature of the primate visual cortex. The HMAX model has four layers: two simple layers (*S*1 and*S*2) and two complex layers (*C*1 and *C*2). In layer *S*1, the input image is densely filtered with Gaussian filters at several orientations and scales. The feature maps generated from layer *S*1 are then arranged into filter bands that contain neighboring scale maps with different orientations. After layer *S*1, the spatial maximum values in each filter band are computed at the same orientation to form layer *C*1. Let PiPi, i=1,...,Ki=1,...,K, be a dictionary that is learned from samples of layer *C*1. PiPi is used as a prototype to represent intermediate-level feature *S*2. The maximum values of *S*2 are computed over all positions and scales, and are used as the output features *C*2, which are shift- and scale-invariant.

The HMAX architecture has the advantages of both template-based features (Viola and Jones 2004) and histogram-based features (Lowe 2004; Dalal and Triggs 2005). HMAX features preserve object geometry similarly to template-based features, and they are robust to small distortions in objects, like histogram-based features. Serre et al. (2007) later used HMAX features for object recognition and scene understanding. In object recognition tasks, the HMAX model outperforms the HOG (Dalal and Triggs 2005), the part-based model (Leibe et al. 2004), and the local patch correlation (Torralba et al. 2004). However, HMAX has a longer processing time than other algorithms, such as the HOG and SIFT.

Several extensions to the HMAX have been proposed. Serre and Riesenhuber (2004) introduced a new HMAX that uses Gabor filters to model simple cell receptive field instead of Gaussian filters. Mutch and Lowe (2006) extended the HMAX by introducing lateral inhibition and feature localization. The extended HMAX achieved an improvement of 14 % in the classification rate compared to the original HMAX, in object categorization on the Caltech-101 data set. Brumby et al. (2009) developed a large-scale functional model based on the HMAX and applied it to detect vehicles in remote-sensing images. Inspired by the HMAX, to describe the scene, Jiang et al. (2010) proposed a new approach that combines features from simple cells and complex cells with the *sparse coding based spatial pyramid matching* (ScSPM) (Yang et al. 2009).

2.1.2 GIST model

Oliva and Torralba (2001) proposed a computational model, known as GIST, for scene categorization. They suggested that images in a scene category possess a similar spatial structure that can be extracted without segmenting the image. The GIST features are the statistical summary of the scene spatial layout. That is, they capture the dominant perceptual properties, such as naturalness, openness, roughness, expansion, and ruggedness, of a scene.

The GIST model can be described as follows. An input image *I* is first padded, whitened, and normalized to reduce the blocking artifact. Next, the image is processed with a set of multi-scale oriented Gabor filters. The impulse response of a Gabor filter is a Gaussian modulated by a harmonic function:

$$g(x, y) = \cos\left(2\pi\frac{x'}{\lambda} + \Phi\right)\ \exp\left(\frac{x'^2 + \gamma^2 y'^2}{2\sigma^2}\right), \tag{1}$$

where $x'=x\cos\theta+y\sin\theta x'=x\cos\theta+y\sin\theta$, $y'=-x\sin\theta+y\cos\theta y'=-x\sin\theta+y\cos\theta$, $\theta\theta$ is the rotation angle, $\Phi\Phi$ is the phase offset, $\lambda\lambda$ is the wavelength of the harmonic function, $\sigma\sigma$ is the standard deviation of the Gaussian function, and $\gamma\gamma$ is the spatial aspect ratio. In the original GIST model, 32 filters at four scales and eight orientations are used. Each output filtered image is partitioned into 16 blocks, and the average value of each block is used as a feature. Overall, a GIST feature vector has 512 elements. Figure 1b shows the GIST features extracted from an input image shown in Fig. 1a. Here, the input image is divided into 4×44×4 regions. From each region 32 gist features (4 scales and 8 orientations) are extracted and visualized in polar coordinates.

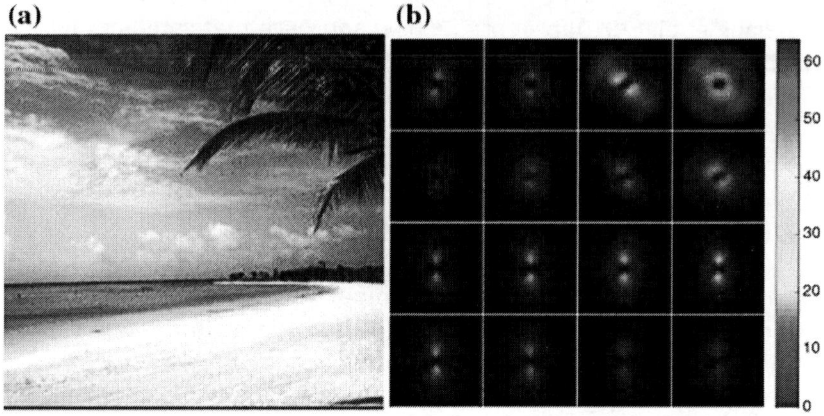

Figure 1. Visual illustration of GIST feature extraction. **a** input image, **b** GIST features in polar form

The GIST features were found to be more effective for recognizing outdoor scenes than the indoor scenes (Wu and Rehg 2011). They have been combined with other features for scene categorization. Torralba et al. (2006) proposed a new GIST model that combines the local features, global features, bottom-up saliency, and top-down mechanisms to predict which image regions are likely to be fixated by human observers. Han and Liu (2010) developed a hierarchical GIST model for scene classification with two layers: (1) a perceptual layer based on the GIST model proposed by Oliva and Torralba (2001); and (2) a conceptual layer based on the kernel PCA (Scholkopf et al. 1998).

2.1.3 Deep learning

In recent years, deep learning architectures have gained fervent research interest for image recognition. One of the major deep learning architectures is the *convolutional neural networks* (CNN) developed by LeCun et al. (1998). CNNs are inspired by the discoveries of Hubel and Wiesel (1968) about the receptive fields in mammal visual cortex. The CNNs are based on three key architectural ideas: (1) *local receptive fields* for extracting local features; (2) *weight sharing* for reducing network complexity; (3) *sub-sampling* for handling local distortions and reducing feature dimensionality. An advantage of CNNs is that they can be trained to map raw pixels to image categories, thereby alleviating the need for hand-designed features.

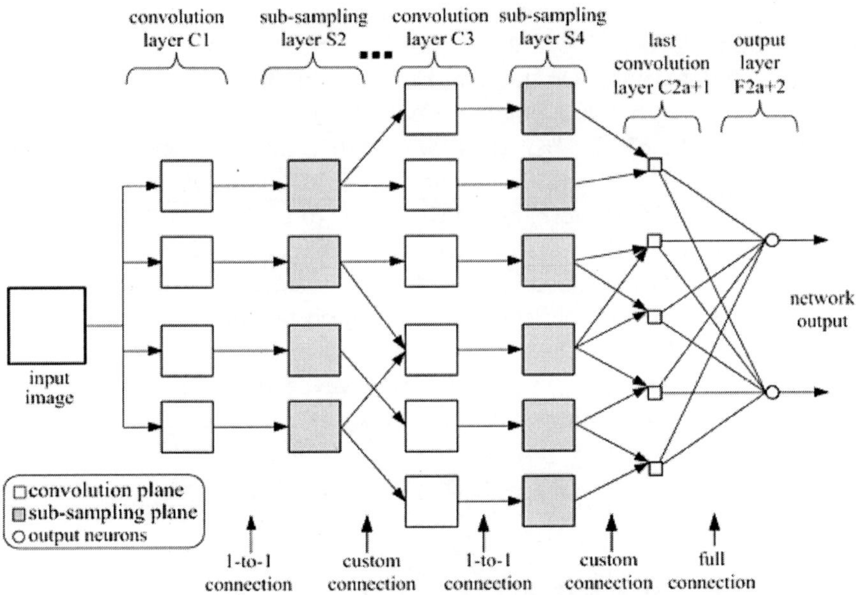

Figure 2. An example of layers in a convolutional neural network

A CNN is a feed-forward architecture with three main types of layers: (1) 2-D convolution layers; (2) 2-D sub-sampling layers; and (3) 1-D output layers (see Fig. 2 for an example). A *convolution layer* consists of several adjustable 2-D filters. The output of each filter is called a feature map, because it indicates the presence of a feature at a given pixel location. A *sub-sampling layer* follows each convolution layer, and reduces the size of each input feature map, via mean pooling or max pooling. The 1-D layers map the extracted 2-D features to the final network output.

Designing and training a CNN or deep learning architecture is a computation-intensive task that requires significant engineering efforts. Hinton et al. (2006) proposed a new approach for training deep networks. Their main idea is to pre-train each layer of the network using an unsupervised learning algorithm, e.g. the*restricted Boltzmann machine*, the *denoising auto-encoder*, and the *kernel principal component analysis*. Once the pre-training is completed, a supervised learning algorithm, e.g. the error backpropagation, is employed to adjust the connection weights of the hidden layers and the output layer (Theodoridis 2015).

Krizhevsky et al. (2012) developed a CNN for image classification, which has 5 convolution layers, 650,000 neurons and 60 million parameters, and produces 4096 features. On the ImageNet benchmark, which comprises 1.2 million images with 1000 object categories, the CNN achieved a top-1 classification rate of 62.5 %, which was higher than the previous results obtained by other methods. Sermanet et al. (2014) later developed a CNN-based integrated framework, called OverFeat, to perform both localization and detection tasks; their system achieved very competitive results (1st in localization and 4th in classification) on the ImageNet benchmark. CNNs have also been applied for scene categorization by Zhou et al. (2014) and Donahue et al. (2014). CNNs have been shown to be less effective for moderate-size data sets (Goh et al.2014), however, they perform well when trained on large-scale data sets (Krizhevsky et al. 2012; Donahue et al. 2014).

2.2 Local feature extraction

Local descriptors capture low-level properties of the scene, whereas global descriptors represent the overall spatial information. Vogel et al. (2007) studied the use of local features and global features in the categorization of natural scenes. Their results suggested that humans rely as much on local region information as on global configural information. Existing algorithms for local descriptors can be divided into three main categories: patch-based, object-based, and region-based.

2.2.1 Patch-based local features

Patch-based algorithms extract features from small patches of input images. For the SIFT (Lowe 2004) and*speeded up robust features* (SURF) (Bay et al. 2008), patches are generated from local windows around interest points. For the LBP (Ojala et al. 1996), patches are formed from rectangular regions of each pixels. For the HOG (Dalal and Triggs 2005), patches are non-overlapping blocks where orientation voting and normalization are applied. For SIFT-ScSPM (Yang et al. 2009), patches are overlapping blocks formed from a regular grid at the same scale. For the SIFT-LCS-FV (Sanchez et al. 2013), patches are overlapping blocks formed from a regular grid at five scales.

(A) Scale-invariant feature transform

Lowe (2004) developed the SIFT algorithm to extract image features that are invariant to image scale, rotation, and changing illumination. Extracting SIFT local features involves four main steps. First, the difference-of-Gaussian filters are applied to identify the location and scale of interest points. Second, the interest points with high stability are selected as the key points. Third, dominant orientation is assigned to each key point based on local image gradient. Fourth, the SIFT features that are partially invariant to affine distortions and illumination changes are extracted from key-point regions. The SIFT features are computed from image gradient magnitude and orientation in a region centered at key point. An example of SIFT key points is shown in Fig. 3a. The centers of circles are the key points, the radiuses of circles are the average scales of the key points, and the arrows inside the circles are the average orientations of the key points.

To apply the SIFT for scene categorization, Fei-Fei and Perona (2005) proposed to extract local features from dense patches. A dictionary is formed from random local patches using *k*-means algorithm. Then, for each input image, a feature vector is generated using the trained dictionary. An example of SIFT feature map that is extracted from dense patches is shown in Fig. 3b. In this figure, the SIFT features are averaged at each pixel location and shown.

The original SIFT algorithm has been extended by several researchers. Brown and Susstrunk (2011) proposed the *multi-spectral SIFT* (MSIFT) on color and near-infrared images for scene categorization. They showed that compared with the SIFT, HMAX, and GIST on the 8-outdoor-scene data set (Oliva and Torralba2001), the MSIFT reduced feature dimensionality and improved recognition accuracy. Liu et al. (2011) proposed the *SIFT flow* to align images across scenes, and applied it for image alignment, video retrieval, face recognition, and motion prediction. Bo et al. (2010) improved the low-level SIFT features using a kernel approach. The kernel descriptors provide a principled tool to convert pixel attributes to patch-level features.

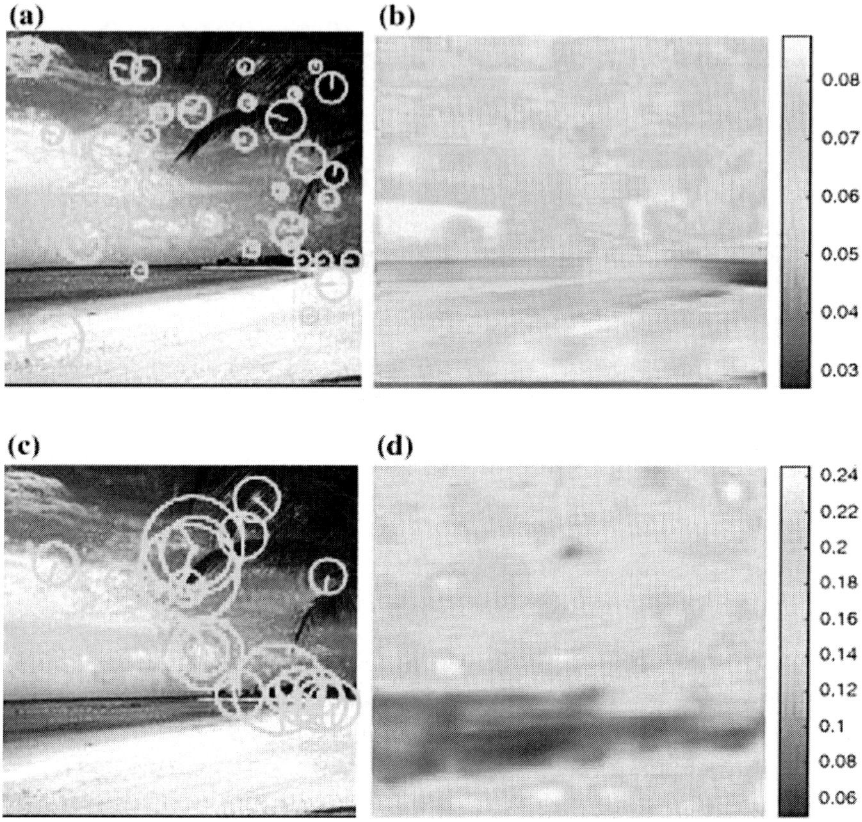

Figure 3. Visual illustration of SIFT, SURF, and HOG feature extraction of the input image in Fig. 1a. **a** SIFT key points, **b** SIFT dense feature map, **c** SURF key points, **d** HOG dense feature map

(B) Speeded-up robust features

Bay et al. (2008) proposed scale- and rotation-invariant visual descriptor, called SURF. The SURF detects interest points using determinant of Hessian matrix. The computation time of interest point detection is reduced by using integral images (Viola and Jones 2001). The key points are then selected from interest points using non-maximum suppression in multi-scale space. The orientation of key point is assigned using sliding orientation windows on Haar wavelet response maps. The longest vector over all windows defines the orientation of the key point. The Haar wavelet response in the horizontal direction (dxdx) and the vertical direction (dydy) are computed from a 4×44×4 sub-region over the key point. The feature vector for each sub-region is v=(∑dx,∑dy,∑|dx|,∑|dy|). SURF key points of an example image is shown in Fig. 3c. The centers of circles are the key points, the radiuses of circles are the average scales of the key points, and the arrows inside the circles are the average orientations of the key points.

(C) Histogram of oriented gradients

Dalal and Triggs (2005) originally developed the HOG descriptor for pedestrian detection in gray-scale images. The HOG features have since been applied for recognition of other image categories, such as cars (Rybski et al. 2010), bicycles (Felzenszwalb et al. 2010), and facial expressions (Bai et al. 2009). The HOG feature extraction involves four main steps. First, an input image is normalized by the power-law, and the image gradients are computed along the horizontal and vertical directions. Next, the image is divided into cells; a cell can be a rectangular or circular region. In the third step, the histograms for multiple orientations are computed for each cell, where each pixel in the cell contributes a weighted score to a histogram. Finally, the cell histograms are normalized and grouped in blocks to form the HOG features. An example of HOG features is shown in Fig. 3d, which illustrates the strength of averaged HOG features at each pixel location.

For scene categorization, the HOG features are useful for capturing the distribution of image gradients and edge directions in a regular grid. Xiao et al. (2014a) compared the HOG with other descriptors, such as GIST and SIFT, on the SUN397 data set. The HOG achieved a higher classification rate (CR) compared with other hand-designed descriptors.

(D) Local binary pattern

Ojala et al. (1996) first developed the LBP algorithm for texture classification. Since then, the LBP has been applied to many computer vision tasks, including face recognition (Ahonen et al. 2006), pedestrian detection (Zheng et al. 2011), and scene categorization (Xiao et al. 2014a). The LBP algorithm analyzes the textures of a local patch by comparing each center pixel with the neighboring pixels.

The basic LBP operates on 3-by-3 blocks. Each pixel in the block is compared to the center pixel, and a binary value of 1 or 0 is returned (see Fig. 4). A pattern B is formed by concatenating the binary values from the neighboring pixels. The decimal LBP code is obtained by summing the thresholded differences weighted by powers of two. Furthermore, a contrast measure *C* is obtained by subtracting the average of pixel values smaller than the center pixel value *p* from the average of pixel values larger than or equal to *p*. An example of LBP code map is shown in Fig. 6. In the example, the pattern *B* is 11001101; the LBP code is 179; the contrast *C* is 7.4. The histogram of local contrast (LBP/CLBP/C) is used as a feature vector. The computational simplicity of the LBP algorithm makes it suitable for real-time image analysis.

input patch (3 by 3 grid)

10	5	2
16	5	1
3	9	7

thresholded

1	1	0
1		0
0	1	1

weights

1	2	4
128		8
64	32	16

pattern B = 11001101 LBP = 1+2+16+32+128 = 179

C = (10+5+7+9+16)/5-(2+1+3)/3 = 7.4

Figure 4. Illustration of the basic LBP algorithm

Several variants of the LBP have been developed. Ojala et al. (2002a) proposed a generic form of LBP that supports arbitrary neighborhood sizes. By contrast to the basic LBP that uses 8 neighboring pixels in a 3-by-3 block, the generic operator LBPP,RLBPP,R is circularly symmetric, see Fig. 5.

Ojala et al. (2002a) also developed the uniform patterns, denoted as LBPuP,RLBPP,Ru. Here, *u* is the number of transitions (0 to 1, or 1 to 0) in an LBP pattern. A local binary pattern is called uniform if u≤2u≤2. Different output labels are assigned to uniform LBP codes, and a single label is assigned to non-uniform patterns. Experiments by Ojala et al. (2002a) indicated that uniform patterns can be considered as fundamental textures because they represent the vast majority of local texture patterns.

Ahonen et al. (2009) proposed an algorithm, called LBP-HF, that combines the uniform LBP and Fourier coefficients. They showed that the LBP-HF has better rotation invariance than the uniform LBP. In the LBP-HF algorithm, the uniform pattern LBPuP,RLBPP,Ru at pixel location(*x*, *y*) is replaced by

LBPuP,(R+θ) mod PLBPP,(R+θ) mod Pu, where θ is a rotation angle: θ=0,2πP,2×2πP,...,(P−1)×2πP. Then, the histograms hθhθ of LBPuP,(R+θ) mod PLBPP,(R+θ) mod Pu are computed. Finally, the Discrete Fourier Transform is applied on hθhθ to form LBP-HF features (Fig. 6).

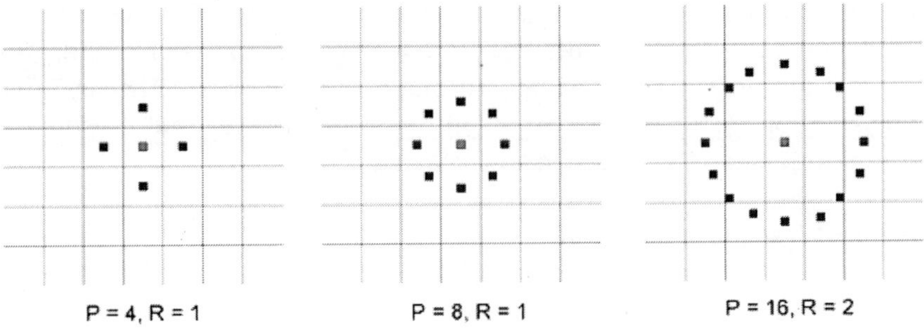

Figure 5. The circular regions in a generic form of the LBP. Here, *P* is the number of neighboring pixels, and *R* is the circle radius. When P=8P=8 and R=1R=1, the basic LBP operator is obtained

Figure 6. Visual illustration of LBP-based feature extraction of the input image in Fig. 1a. **a** LBP, **b** uniform LBP, **c**LBP-HF

Guo et al. (2010) proposed a *completed local binary pattern* (CLBP) to extract image features for texture classification. The original LBP only encodes the signs of differences between center pixel and its neighbors (see Fig. 4). The CLBP encodes both the signs (CLBP-S) and magnitudes (CLBP-M) of the differences. Furthermore, the intensity of center pixel (CLBP-C) is encoded as the third part of CLBP. Guo et al. (2010) showed that the texture classification accuracy of CLBP was better than the original LBP algorithm. Li et al. (2012) proposed a scale- and rotation-invariant LBP descriptor. The scale-invariance of this method is achieved by searching for the maximum response over scale spaces. The rotation invariance is achieved by locating the dominant orientations of the uniform-LBP patterns. The scale- and rotation-invariant LBP outperforms the classical uniform LBP in texture classification. In another approach, Qian et al. (2011) proposed an LBP descriptor with pyramid representation (PLBP). By cascading the LBP features obtained from hierarchical spatial pyramids, the PLBP descriptor extracts texture resolution information. The pyramid representation for LBP is more efficient than the multi-resolution representation for LBP proposed by Ojala et al. (2002b).

(E) Census transform histogram

Wu and Rehg (2011) proposed a visual descriptor called CENTRIST, which is a holistic representation of structural and geometrical properties of images. In the CENTRIST, feature maps are calculated by the census transform (CT), which is equivalent to the local binary pattern LBP8,1LBP8,1. Wu and Regh presented an experiment showing that CT values encode shape information. The authors identified 6 CT values with highest counts (31, 248 240, 232, 15, and 23) in the 15-scene data set (Fei-Fei and Perona 2005; Lazebnik et al. 2006). The 6 CT values correspond to local 3-by-3 neighborhoods that have horizontal or close-to-diagonal edge structures. To encode the global structure of an image, the CT values are processed by the spatial pyramid algorithm, described in (Lazebnik et al. 2006). The features extracted from the pyramid feature maps are then reduced using the spatial PCA or BoW methods.

Wu and Rehg (2011) showed experimentally that the CENTRIST with the spatial PCA outperforms the state-of-the-art algorithms, such as the SIFT and GIST, on several scene categorization data sets. However, the CENTRIST has a number of limitations. First, the CENTRIST is not invariant to rotations. Second, it is not designed for extracting color information. Third, the CENTRIST still has difficulty in learning semantic concepts from images with varied viewing angles and scales. Based on the CENTRIST, a multi-channel feature generation mechanism was proposed by Xiao et al. (2014b). The CENTRIST features with multi-channel information (RGB channels and infrared channel) improves grayscale CENTRIST's performance on scene categorization.

2.2.2 Object-based local features

Object-based algorithms rely on landmark objects to classify scenes. They have been applied for scene perception in robotic navigation systems (Abe et al. 1999; Yu et al. 2009b; Schauerte et al. 2011). Bao et al. (2011) proposed a scene layout reconstruction algorithm by determining the 3-D locations and the support planes of objects. In these methods, the scene is classified based on landmark objects and their configuration. A challenge of object-based methods is to detect small objects, especially in outdoor conditions. Another challenge is to select a small set of landmark objects to represent a scene (Siagian and Itti 2007).

An example of object-based algorithms for visual categorization are *object bank* proposed by Li et al. (2010). The OB algorithm was designed to decrease the gap between low-level visual features of objects and high-level semantic information of the scene.

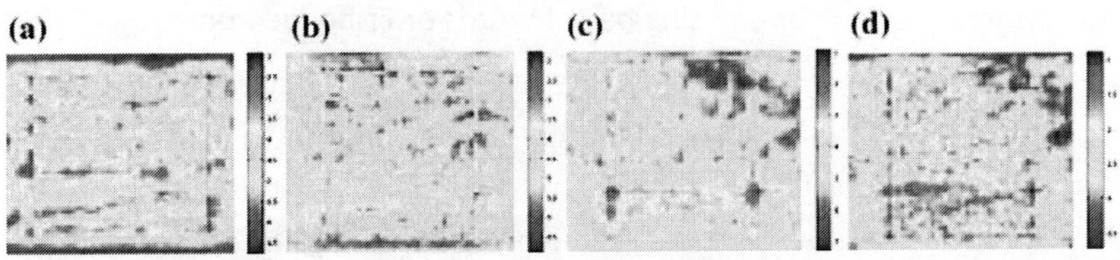

Figure 7. Object bank feature maps of input image in Fig. 1a. **a** 'tree', **b** 'sky', **c** 'beach', **d** 'coast'

In the OB approach, an image is represented by scale-invariant response maps produced by pre-trained object detectors. Objects are classified by two types of detectors: (1) the SVM detector, proposed by Felzenszwalb et al. (2010), for objects such as humans, cars, and tables; (2) the texture detector, proposed by Hoiem et al. (2006), for objects such as sky, road, and sea. Li et al. (2010) analyzed common object types in four data sets ESP (Von Ahn 2006), LabelMe (Russell et al. 2008), ImageNet (Jia et al. 2009), and Flickr (Yahoo 2004), and selected 200 object detectors that were trained with 12 detection scales and 3 spatial pyramid levels (Peters and Itti 2007). The example feature maps produced by four object detectors are shown in Fig. 7a–d. Li et al. (2010) compared the OB algorithm with SIFT, GIST, and *spatial pyramid matching* (SPM) (Lazebnik et al. 2006) on several scene data sets. The results showed that OB outperforms the other algorithms on the UIUC data set (Li and Fei-Fei 2007), the 67-indoor-scene data set (Quattoni and Torralba 2009), and the 15-scene data set (Lazebnik et al. 2006).

2.2.3 Region-based local features

Region-based algorithms segment images and extract features from different regions. Boutell et al. (2006) proposed an algorithm to classify scenes using the identities of regions and the spatial relations between regions. Gokälp and Aksoy (2007) proposed a bag-of-regions algorithm for scene classification. In their algorithm, an image is partitioned into regions, and the structure of the image is represented by a bag of individual regions and a bag of region pairs. Juneja et al. (2013) used distinctive parts for scene categorization. First, the distinctive parts in each category are detected and learned based on HOG features. Then, the features of parts are extracted and encoded using bag of words or Fisher vector.

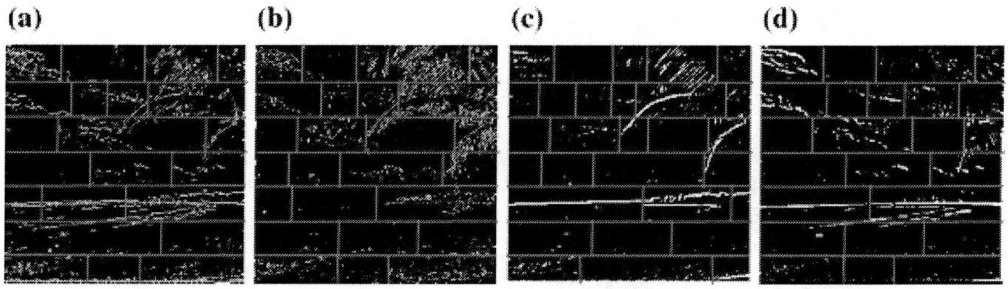

Figure 8. Illustration of SEV regions of input image in Fig. 1a and the computed edge maps along the **a** horizontal direction, **b** vertical direction, **c** +45+45 degree direction, and **d** −45−45 degree direction

An example of region-based algorithms is the *sequential edge vectors* (SEV) proposed by Morikawa and Shibata (2012). The SEV reduces the ambiguity of features caused by the inter-class variations among scene categories. Unlike other region-based algorithms that segment the entire image, the SEV segments only local regions. The SEV method considers an image as a *document* consisting of *sentences*, in which the separated regions play the role of *words*, and oriented edges play the role of *letters*. The SEV identifies each word from the letters, forms the document vector from words, and finally determines the topic vector.

The main steps of the SEV method can be described as follows. First, oriented edges of input image are detected by horizontal, vertical, and diagonal edge filters. Second, the horizontal or vertical edge map is used to generate fixed sentences (also called local threads). All local threads are scanned from one end to the other, producing edge distribution vectors in four directions. Next, the sum of absolute differences (SAD) between neighboring local windows is calculated. The boundaries of meaningful sequences are determined by locating peaks in the SAD histograms. Figure 8 shows the boundaries in four edge maps. Then, the meaningful sequences form words that describe images. Finally, probabilistic latent semantic analysis (PLSA) (Quelhas et al. 2005) is applied on meaningful sequences (words) to generate topic vectors.

Using the 8-outdoor-scene data set, Morikawa and Shibata evaluated the SEV algorithm on seven scene categories: coast, open country, forest, highway, mountain, street, and tall building. The SEV method with a 16-by-16 scan window achieved a higher *F*-measure (65 %) than a model that uses SIFT and PLSA (55 %).

2.3 Global feature formation

For scene categorization, global features are often extracted without image segmentation or object detection to summarize the statistics of the image. For example, Renninger and Malik (2003) computed histograms of features generated by a bank of Gaussian derivative filters to represent scenes. Serrano et al. (2004) used quantized color histograms and wavelet texture as global features for scene categorization. Furthermore, Mikolajczyk and Schmid (2005) showed that the performance of visual descriptors depends not only on local regions but also global information. Therefore, several scene categorization algorithms focus on first extracting suitable local features and then forming global features from regular grids. In the following subsections, four representative algorithms for global feature formation in scene categorization are described: the PCA, histograms, BoW, and Fisher vector.

2.3.1 Principal component analysis

Principal component analysis (PCA) represents data by a small number of principal components. It has been used for dimensionality reduction in a wide range of computer vision tasks, such as image categorization (Han and Chen 2009), face recognition (Guangda et al. 2002; Xie and Lam 2006), and feature selection (Malhi and Gao 2004; Yang and Li 2010; Ebied 2012).

Ke and Sukthankar (2004) proposed a visual descriptor, known as PCA-SIFT, that combines the SIFT and PCA. In their approach, a projection matrix P is calculated from a large number of image patches. For each training image patch, a feature vector is generated from the horizontal and vertical gradient maps. The covariance matrix C of all feature vectors is calculated. Eigenanlysis is applied to C, and the n most-significant eigenvectors of C are selected to form the projection matrix P. In Ke and Sukthankar's experiments, n was selected to be 20, which is significantly smaller than the number of features (128) in the standard SIFT algorithm. Compared to the SIFT features, the PCA-SIFT features lead to an improvement in image matching, when evaluated on the INRIA Graffiti data set (2004).

2.3.2 Histogram

Histogram is a method to represent the statistical distribution of data. It is efficient and robust to noise compared to other feature formation methods (Hadjidemetriou et al. 2004). Therefore, it has been used in many image processing tasks, including image and video retrieval (Lee et al. 2003; Song et al. 2004; Jeong et al. 2003; Liu and Yang 2013), image structure analysis (Koenderink and Van Doorn 1999), image filtering (Kass and Solomon 2010; Igarashi et al. 2013), and color indexing (Swain and Ballard 1991; Stricker and Orengo 1995; Funt and Finlayson 1995; Kikuchi et al. 2013). In several scene categorization algorithms, such as the LBP and its variants, global features are formed by calculating the histogram of local features (Ojala et al. 1996; Gupta et al. 2009; Wu and Rehg 2011).

A weakness of the histogram approach is that it does not capture the spatial information. To encode spatial information, Hadjidemetriou et al. (2004) proposed multi-resolution histograms that extract shape and texture features at several resolution levels. Given an image $I(x, y)$ with n gray-levels, the spatial resolution of the image is decreased by convolving it with a Gaussian kernel $G(x,y;\sigma)G(x,y;\sigma)$. The multi-resolution histograms are calculated as h[I∗G(σ)]h[I∗G(σ)]. Then, the cumulative histograms corresponding to each image resolution is computed. Next, the differences between the cumulative histograms of consecutive levels are calculated. The difference histograms are sub-sampled and normalized to make them independent of the sub-sampling factor. Finally, the normalized difference histograms are concatenated to form a feature vector.

2.3.3 Bag-of-words

The *bag-of-word* (BoW) algorithms were first used in document classification to simplify the representation of natural language. Recently, BoW has been used to classify images based on the appearance of image patches. In the BoW, features are extracted from regular grids (*feature extraction*) and quantized into discrete visual words (*encoding*). The coding strategy aims to generate similar codes for similar features. A compact representation of visual words is built for each image based on a trained dictionary (*codebook*).

Qin and Yung (2010) proposed an algorithm based on contextual visual words. To train the visual words, SIFT features are calculated from both the *region of interest* (ROI) and the regions surrounding the ROI. Fei-Fei and Perona (2005) proposed a BoW algorithm based on *latent Dirichlet allocation* (Blei et al. 2003). They compared different local patch detectors, such as the regular grids, random sampling, saliency detector (Kadir and Brady 2001), and the difference-of-Gaussian detector (Lowe 2004). Their experiment results showed that the regular grids perform better than the random sampling, saliency detector, and difference-of-Gaussian detector for scene categorization.

Next, we describe three representative BoW algorithms for training visual words on regular grids: SPM (Lazebnik et al. 2006), ScSPM (Yang et al. 2009), and LLC (Wang et al. 2010).

Spatial pyramid matching (SPM) for recognizing natural scenes was proposed by Lazebnik et al. (2006). Unlike the traditional BoW algorithms that extract order-less features, the SPM algorithm retains the global geometric correspondence of images. It divides the input image into regular grids and computes local features, such as SIFT and HOG in each grid. The visual vocabulary is formed by *k*-means clustering, and then all features are formed using vector quantization (VQ). Based on the trained dictionary, local features are represented. Finally, the spatial histograms (average pooling) of coded features are used as feature vectors. To recognize multiple scene categories, a support vector machine with the one-versus-all strategy is used.

Sparse coding based spatial pyramid matching (ScSPM) was suggested by Yang et al. (2009) to improve the efficiency of the SPM. The original SPM uses *k*-means vector quantization, whereas the ScSPM uses sparse coding to quantize the local features. Furthermore, for spatial pooling, the original SPM uses histograms, whereas the ScSPM applies the *max* operator, which is more robust to local spatial translations. For the ScSPM, the linear-SVM classifier is used to reduce the computation cost. The experiments by Yang*et al.* show that the sparse coding of SIFT descriptors with the linear-SVM outperforms several methods, including kernel codebooks (Gemert et al. 2008), SVM-K-Nearest-Neighbor (Zhang et al. 2006), and naive Bayes nearest-neighbor (NBNN) (Boiman et al. 2008).

Locality-constrained linear coding (LLC) was proposed by Wang et al. (2010) to reduce the computational cost of the SPM. Because of the importance of locality, as demonstrated by Yu et al. (2009a), the LLC replaces the sparsity constraint used in ScSPM with the locality constraint to select similar bases for local features from the trained visual words. In Wang et al.'s approach, a linear weighted combination of these bases is learned to represent local features. Their experiments on the Catech-101 data set (Fei-Fei et al.2007a) show that the LLC achieved higher classification rates than the ScSPM, NBNN (Boiman et al. 2008), and kernel codebooks (Gemert et al. 2008). Recently, Goh et al. (2014) improved the ScSPM and LLC by using a deep architecture. The deep architecture merges the strengths of the BoW framework and the deep learning method to encode the SIFT features.

2.3.4 Fisher vector

Fisher vector (FV) is a feature encoding algorithm proposed by Sanchez et al. (2013). The local features extracted from dense multi-scale grids are represented by their deviation from a *Gaussian mixture model*(GMM). The local features are mapped to a higher-dimensional space which is more amenable to linear classification.

In the FV algorithm, a GMM is first computed from a training set of local features using the Expectation–Maximization algorithm. The parameters of a GMM are denoted by λ={(wk,μμk,σσk),k=1,...,K}λ={(wk,μμk,σσk),k=1,...,K}, where wkwk is the mixture weight, μμkμμk is the mean vector, and σσkσσk is the covariance matrix of the *k*th Gaussian component.

Let X={xt,t=1,...,T} be the set of local descriptors extracted from an input image. For local descriptor xtxt, let pλ(xt) be the probability density function of local descriptor, as computed by the GMM model. Let Lλ be the square-root of the inverse of the Fisher information matrix. The normalized gradient statistics are computed as

$$\varphi(\mathbf{x}_t) = L_\lambda \nabla_\lambda \log p_\lambda(\mathbf{x}_t). \tag{2}$$

A Fisher vector is the sum of normalized gradient statistics:

$$\mathcal{G}_\lambda^X = \sum_{t=1}^{T} \varphi(\mathbf{x}_t) \tag{3}$$

The final Fisher vector is the concatenation of the gradients GXwk, GXμμkGμμkX, and GXσσk. To improve the classification accuracy, two normalization steps, l2l2-normalization and power normalization (Perronnin et al.2010), can be applied on the Fisher vector.

Compared with other bag-of-words algorithms, the Fisher vector has many advantages. First, it provides a generalized method to define a kernel from a generative process of the data. Second, Fisher vector can be computed from small vocabularies with a lower computational cost. However, the Fisher vector is dense, which leads to storage issues for large-scale applications.

2.3.5 Composite global features

Global features are also formed by combining different global features. For example, the CENTRIST features are generated by combining histograms and PCA (Wu and Rehg 2011). The HOG-SPM and SIFT-SPM features are first extracted by BoW algorithms, and then accumulated by spatial histograms (Lazebnik et al.2006). The SIFT-ScSPM (Yang et al. 2009) and SIFT-LLC (Wang et al. 2010) represent local features by the BoW algorithm and the max pooling. The image categorization method proposed by Krapac et al. (2011) combines the SPM and the Fisher kernel to encode spatial layout of images.

3. DATA SETS AND PERFORMANCE MEASURES

Progress in image recognition is due in part to the existence of comprehensive data sets on which new or existing algorithms can be rigorously evaluated (Salzberg 1997; Phillips et al. 2000). In this section, we review the publicly available data sets and the performance measures for scene categorization algorithms, and discuss their characteristics.

3.1 Data sets for scene categorization

Because of the difficulty of finding one representative data set, it is not sufficient to evaluate a scene categorization algorithm on only one data set. For example, on the Caltech-101 data set, the LLC algorithm is found to have a higher classification rate than the ScSPM algorithm (Wang et al. 2010). However, on the 15-scene data set, the ScSPM algorithm has a higher classification rate than the LLC algorithm (see Sect. 4.4). One possible reason is that built-in biases are present when collecting image data for a recognition task, e.g. the viewing angle, the type of background scene, and the composition of objects. These intrinsic biases cause every data set to represent the physical world differently. Therefore, scene categorization algorithms should be evaluated on multiple large-scale data sets that exhibit more diversity and less bias.

Table 3 summarizes the major data sets used for scene categorization. For each data set, the source, the number of images, and the number of image categories are given. In the following, we describe the major data sets. Three benchmark data sets, 8-outdoor-scene (Oliva and Torralba 2001), 13-natural-scene (Fei-Fei and Perona 2005), and 15-scene (Lazebnik et al. 2006), have been used by many researchers (Grossberg and Huang 2009; Qin and Yung 2010; Jiang et al. 2010; Kazuhiro 2011; Cakir et al. 2011; Morikawa and Shibata 2012; Meng et al. 2012; Goh et al. 2014; Zhang et al. 2015). The 8-outdoor-scene data set has eight categories of only-outdoor scenes (coast, forest, highway, inside city, mountain, open country, street, and tall buildings), whereas the 13-natural-scene data set contains the same eight categories of outdoor scenes and five additional categories of indoor scenes (bedroom, kitchen, living-room, office, and store). The 15-scene data set includes all images from the 13-natural-scene, and two additional outdoor scenes of man-made structures (suburb and industry). The images in each category are from different sources, such as COREL data set, Google image search, and personal photographs.

Large data sets have been developed to increase the number of semantic categories and the diversity of images. The 67-indoor-scene data set (Quattoni and Torralba 2009) divides five indoor scenes (working places, home, leisure, store, and public spaces) into 67 sub-categories. Common objects appear in multiple categories; therefore, it is harder to distinguish between the image categories. The SUN397 data set (Xiao et al. 2014a) has 397 scene categories, from abbey, bedroom, and castle to highway, theater, and yard. There are at least 100 images in each category. The Places205 data set (Zhou et al. 2014) has 205 scene categories; each category has at least 5000 images. Among the benchmark data sets listed in Table 3, the SUN397 has the most number of categories and the Places205 has the most number of images.

3.2 Performance measures

To evaluate a scene categorization system, a data set is typically partitioned into three separate subsets: training, validation, and test. The training subset is used to determine the system's adjustable parameters, the validation set is used to prevent over-training, and the test set is used to estimate the system's generalization capability. The generalization capability of a system is commonly measured using the classification rate (*CR*), which is the percentage of test images that are correctly classified. For example, the *CR* has been used for scene categorization in (Elfiky et al. 2012; Wu and Rehg 2011; Kazuhiro 2011; Meng et al. 2012).

To prevent bias in partitioning the data set and to estimate more reliably the generalization capability, an alternative technique known as *n*-fold *cross-validation* is usually adopted. The image data set is divided into*n* subsets of equal size. For each validation fold, one subset is used for testing, and the remaining (n−1)(n−1)subsets are used for training and validation. This is repeated *n* times, each time a different subset fold is used for testing. Finally, the *n* classification rates are averaged to give the overall *CR*.

Table 3. Data sets for scene categorization

Data set	Data set size (images)	Image categories
8-outdoor-scene (Oliva and Torralba 2001)	2600	8 outdoor scenes
13-natural-scene (Fei-Fei and Perona 2005)	3759	13 natural scenes
15-scene (Lazebnik et al. 2006)	4485	15 indoor/outdoor scenes
67-indoor-scene (Quattoni and Torralba 2009)	15,620	67 indoor scenes
SUN397 (Xiao et al. 2014a)	108,754	397 general scenes
Places205 (Zhou et al. 2014)	2,448,873	205 general scenes

Scene categorization is a multi-class recognition problem. Many scene categorization algorithms are also evaluated using the *confusion matrix* (Wu and Rehg 2011; Perina et al. 2010; Meng et al. 2012; Morikawa and Shibata 2012; Cakir et al. 2011; Qin and Yung 2010). For a problem involving *K* categories, the confusion matrix has *K* rows and *K* columns. Each row represents an actual category, and each column represents a predicted category. The entry at row *r*, column *c* is the number of category *r* samples that are classified as category *c*. Clearly, the correct classification for individual categories are the diagonal entries, whereas the miss-classification are the non-diagonal entries.

Evaluation measures for two-class problems are also applied for scene categorization. Consider a scene category *r*. The positive class consists of all samples belonging to category *r*, whereas the negative class consists of all samples belonging to the remaining categories. Four measures can be computed:

- *True positives* (*tp*) is the number of test samples in the positive class that are correctly classified.
- *False positives* (*fp*) is the number of test samples in the negative class that are incorrectly classified.
- *False negatives* (*fn*) is the number of test samples in the positive class that are incorrectly classified.
- *True negatives* (*tn*) is the number of test samples in the negative class that are correctly classified.

The precision rate *P* and the recall rate *R* are then defined as

$$P = \frac{tp}{tp + fp}, \text{ and } R = \frac{tp}{tp + fn}. \tag{4}$$

A good scene categorization system should have a high precision rate and a high recall rate. These two requirements can be reflected in a single measure called the *F-measure*, which is the harmonic mean of the precision rate and recall rate:

$$F = \frac{2PR}{P + R}. \tag{5}$$

The plot of the true positive rate versus the false positive rate is called the *receiver operating characteristic*(*ROC*) curve (Fawcett 2006). It is a useful tool for visualizing the scene categorization performance, when a system parameter is varied. Another performance measure is the *area-under-the-ROC-curve* or *AUC*. The*ROC* curve and the *AUC* have been used for scene categorization in (Jia et al. 2009) and (Xiao et al. 2014a).

The measures described above are suitable for evaluating the performance of a complete scene categorization system that includes both a feature extractor and a classifier. To evaluate the performance of the feature extraction independently of the classifier, we can use the class separability of the extracted features. The *Fisher's discriminant analysis* (FDA) is a tool for analyzing the separability of features. For example, Li et al. (2003) and Chin et al. (2011) used FDA to analyze multi-class image classification. Consider a scene categorization problem that involves *K* classes. The within-class covariance matrix CwCw is calculated as

$$C_w = \sum_{k=1}^{K} \sum_{\mathbf{x} \in \omega_k} (\mathbf{x} - \bar{\mathbf{x}}_k)(\mathbf{x} - \bar{\mathbf{x}}_k)^T, \tag{6}$$

where x¯k is the mean vector of class ωk. The between-class covariance matrix CbCb is given by

$$C_b = \sum_{k=1}^{K} N_k (\bar{\mathbf{x}}_k - \bar{\mathbf{x}})(\bar{\mathbf{x}}_k - \bar{\mathbf{x}})^T, \tag{7}$$

where x¯ is the mean vector of all classes and NkNk is the number of samples in class ωkωk. The *S score* of the feature vector is given by

$$S = \frac{|\mathrm{trace}(C_b)|}{|\mathrm{trace}(C_w)|}. \tag{8}$$

A high *S* score means there is a high separability between the *K* classes using the given feature vector.

4. EXPERIMENTAL EVALUATION AND RESULTS

In this section, we present an extensive experimental evaluation of the performance of state-of-the-art descriptors, which include biologically-inspired, local, and global feature extraction methods. The selected descriptors were evaluated on four benchmark data sets with respect to scene categorization accuracy and class separability of feature vectors. For scene categorization, five measures are used to evaluate the classification accuracy, namely classification rate, precision, recall, F-measure, and *AUC*. By using the same classification protocol described in (Krizhevsky et al. 2012), we are able to compare descriptors on the SUN397 data set with several recent methods: ImageNet-CNN (Krizhevsky et al. 2012), BOP-FV (Juneja et al. 2013), discriminative patches (Doersch et al. 2013), OverFeat (Sermanet et al. 2014), Places-CNN (Zhou et al. 2014), and DeCAF (Donahue et al. 2014). For class separability, the visual descriptors are compared using Fisher discriminant analysis.

Section 4.1 describes the four image data sets used in the experiments. Section 4.2 and 4.3 describe the implementation of the descriptors and classifiers. Section 4.4 presents results of a comparative study using different classifiers with four data sets. Finally, Sect. 4.5 evaluates the class separability and stability of feature vectors.

4.1 Image data sets

The experiments were conducted on four data sets: the 8-outdoor-scene data set, the 15-scene data set, the 67-indoor-scene data set, and the SUN397 data set. The 8-outdoor-scene data set and the 15-scene data set have been used as benchmark for scene categorization by many researchers (Fei-Fei and Perona 2005; Yang et al. 2009; Quattoni and Torralba 2009; Li et al. 2010; Jiang et al. 2010; Wu and Rehg 2011; Kazuhiro2011; Karayev et al. 2014). The 67-indoor-scene data set contains 67 indoor scene categories. There are at least 100 images per category and all images have a size of at least 200×200200×200 pixels. The 67-indoor-scene data set has been used to evaluate scene categorization in (Quattoni and Torralba 2009; Wang et al. 2010; Wu and Rehg 2011; Doersch et al. 2013; Juneja et al. 2013; Bergamo and Torresani 2014; Cheng et al. 2015; Dixit et al. 2015). The SUN397 data set contains 397 scene categories and 108,754 images. It has been used as a benchmark for scene categorization by Huang et al. (2014b), Sanchez et al. (2013), Sun et al. (2013), Bergamo and Torresani (2014), Zhou et al. (2014), Donahue et al. (2014), and Dixit et al. (2015).

4.2 Implementation of visual descriptors

This section describes the implementation of the visual descriptors compared in our experiments. There are two biologically-inspired descriptors (GIST and HMAX), four SIFT-based descriptors (SIFT-SPM, SIFT-ScSPM, SIFT-LLC, and SIFT-FV), two other BOW-based descriptors (HOG-SPM and SURF-ScSPM), five LBP-based descriptors (LBP, Uniform LBP, LBP-HF, PLBP, CENTRIST), and one object-based descriptor (OB). Most of the 14 descriptors combine local and global features.

The biologically-inspired descriptors implemented in this paper are GIST and HMAX. The *GIST descriptor* is a low dimensional representation of an image. In our experiment, the normalized input image was convolved with Gabor filters at 4 scales and 8 orientations. Each filtered output was down-sampled to a 4 by 4 patch and reshaped to a 16 element vector. The GIST descriptor assembled all outputs from the 32 filters to form a feature vector with 512 elements. The *HMAX descriptor* contains two simple layer $S1$ and $S2$, and two complex layer $C1$ and $C2$. In our experiment, the $S1$ layer was formed from the outputs of Gaussian filters with 4 orientations and 12 scales. In layer $C1$, the $S1$ feature maps were grouped into 4 filter bands of a certain scale range. The max pooling operation was applied to each filter bank. Only $S1$ units with the same preferred orientation fed into a given C1 unit. The $S2$ features were formed from $C1$ features with 256 visual words (learned from $C1$). The $C2$ features were generated from $S2$ features using max pooling. The final HMAX descriptor had 4069 features.

The SIFT-based descriptors were compared to evaluate the encoding capability of SPM, ScSPM, LLC, and FV. For the four SIFT-based descriptors (SIFT-SPM, SIFT-ScSPM, SIFT-LLC, and SIFT-FV), their local features were calculated from overlapping patches (16×1616×16) on a dense grid every 8 pixels. The local patch was first filtered with Gaussian filters to generate 8 orientation maps. The histograms were then generated from the orientation maps and further weighted by a Gaussian function. The SIFT descriptor was formed by concatenating the orientation histograms. The *SIFT-SPM descriptor* extracted global features from dense SIFT features by the SPM algorithm. In the experiments, *k*-means clustering and PCA were used to train and extract 200 visual words from random samples of local features. The local features were quantized by the trained visual words. Finally, histograms of quantized features were formed with 1000 bins. The *SIFT-ScSPM descriptor* formed global features from dense-SIFT features using the ScSPM algorithm. A feature dictionary including 1024 visual words was obtained by applying *k*-means clustering to the local features. Finally, the ScSPM algorithm was employed to convert the local SIFT features to global features. The *SIFT-LLC descriptor* formed global features from dense-SIFT features using the LLC algorithm. The SIFT-LLC descriptor converted the local feature maps to global features based on the *n* nearest neighbors of the feature dictionary. In the experiment, the number of neighbors was set to 5. The *SIFT-FV descriptor*extracted global features from dense-SIFT features using the Fisher vector encoding. The dictionary was trained using a Gaussian mixture model with 256 Gaussians.

Other BOW-based descriptors that are similar to the SIFT-based descriptors include HOG-SMP and SURF-ScSPM. In our experiment, the *HOG-SPM descriptor* extracted local features from overlapping patches (16×1616×16) on a dense grid every 8 pixels. The input image was first normalized globally by the power-law. Then, image gradients were computed along the horizontal and vertical directions. For each patch, histograms for multiple orientations were computed to form HOG features. The global features were then generated by the spatial pyramid matching (SPM). The *SURF-ScSPM descriptor* used the SURF for local feature extraction. The ScSPM algorithm was applied for global feature formation. Different from the SIFT-based descriptors and HOG-SMP that extracted local features from dense patches, the SURF-ScSPM descriptor extracted local features from interest-points (sparse patches). In our experiment, at least 100 interest points were found in each image.

The LBP-based descriptors compared in the experiment include LBP, uniform LBP, LBP-HF, PLBP, and CENTRIST. The *LBP descriptor* is a histogram of the LBP feature map. In our experiment, the feature map was generated from 3-by-3 blocks of the entire image. The number of histogram bins was selected as 256. When the number of histogram bins was reduced, our preliminary experiments indicated that the classification rate dropped by about 10 %. The *uniform LBP descriptor* is similar to the LBP descriptor, but it only encodes uniform patterns of the LBP. For the uniform LBP, we represented each input image with 59 uniform patterns. The histograms of the uniform LBP were calculated with 59 bins. The *LBP-HF descriptor* is an extension of the uniform LBP. In our experiment, the histograms of the uniform LBP were first calculated on the input image and its 90-degree rotated version. Then, the Fourier transform was applied on the histograms. The magnitudes of Fourier coefficients were calculated as the LBP-HF features. The final LBP-HF vector had 76 elements: half of the elements were generated from the original image, and the other half from the rotated image. The *PLBP descriptor* is an LBP descriptor with spatial pyramid representation. To calculate PLBP features, each input image was decomposed into 5 Gaussian pyramid images with dyadic scales. The histograms of the LBP in each pyramid image were combined to form the PLBP features. The*CENTRIST descriptor* first converted input image into the CENTRIST feature map (similar to LBP feature map). Then, spatial histogram with 3 spatial levels and PCA with 40 eigenvectors were employed to form the CENTRIST feature map. The dimension of the CENTRIST was 1240.

The object-based descriptor OB extracts features from a large number of pre-trained object detectors. It has the longest feature-extraction stage among the compared descriptors. In our experiment, 176 object detectors with 12 detection scales and 3 spatial pyramid levels were used. An OB descriptor with 44,604 features was formed by max pooling.

4.3 Classification procedure

Support Vector Machines (SVM) with linear, RBF and HIK kernels were used to classify the different descriptors. The linear kernel is given by

$$K(\mathbf{f}_i, \mathbf{f}_j) = \mathbf{f}_i \cdot \mathbf{f}_j, \tag{9}$$

where fi and fj are two feature vectors. The RBF kernel is given by

$$K(\mathbf{f}_i, \mathbf{f}_j) = \exp\left\{-\gamma ||\mathbf{f}_i - \mathbf{f}_j||^2\right\}, \tag{10}$$

where γ is a positive scalar. The HIK kernel is computed as

$$K(\mathbf{f}_i, \mathbf{f}_j) = \sum_{n=0}^{N} \min\left[\mathbf{h}_i(n), \mathbf{h}_j(n)\right], \tag{11}$$

where hi(n) and hj(n)h are, respectively, the *N*-bin histograms of fi and fj.

Five-fold cross validation was used to evaluate the performance of the SVM classifiers with different visual descriptors. In each fold, four subsets were used for training and validation, and the remaining subset was used for testing. The parameters of the SVM classifiers were determined using a validation set in each fold. The average values of *CR, P, R, F,* and *AUC* were calculated over the five folds. The standard deviations of*CR, P, R, F,* and *AUC* over the five folds are used as a measure of variation in the classification performance.

4.4 Classification results

The following four subsections present and discuss the results of image classification on the 15-scene, 8-scene, 67-indoor-scene, and SUN397 data sets.

4.4.1 Classification results on the 15-scene data set

In the first experiment, the 14 selected descriptors were evaluated on the 15-scene data set. Tables 4, 5 and6 present the classification performance measures of the different visual descriptors using the linear-SVM, RBF-SVM, and HIK-SVM. In these tables (and also Tables 7, 8, and 9), the best performance measure is indicated in bold font. The results in these tables show that each descriptor achieves its best classification rate using a different SVM classifier. For example, the SIFT-ScSPM, SIFT-LLC, SIFT-FV, SURF-ScSPM, CENTRIST, and OB achieve higher classification rates with the linear-SVM than with the RBF- or HIK-SVM. The LBP-HF and GIST have their highest classification rates when using the RBF-SVM. The HOG-SPM, SIFT-SPM, LBP, uniform LBP, PLBP, and HMAX achieve their highest classification rates with the HIK-SVM. The highest classification rates on the 15-scene data set for individual descriptors are (in a descending order) the SIFT-ScSPM (84.5 %), SIFT-LLC (83.0 %), SIFT-FV (80.2 %), HOG-SPM (80.0 %), OB (79.9 %), SIFT-SPM (77.9 %), SURF-ScSPM (73.3 %), PLBP (73.0 %), GIST (72.8 %), CENTRIST (72.7 %), LBP (71.9 %), uniform LBP (70.6 %), LBP-HF (67.3 %), and HMAX (63.9 %).

The biologically-inspired algorithms HMAX algorithm has the lowest *CRs* among the compared algorithms. As shown in Tables 4, 5 and 6, the *CRs* of the HMAX using the linear-SVM, RBF-SVM, and HIK-SVM are 61.1, 62.4, and

63.9 %, respectively. The other biologically-inspired descriptor, namely the GIST, has *CRs* of 71.5, 72.8, and 72.1 %, respectively. Furthermore, the GIST algorithm performed better than the LBP-based and HMAX algorithms with all three SVM kernels.

The SIFT-ScSPM outperforms all other 13 descriptors on the 15-scene data set; its *CRs* is 84.5 % for the linear-SVM, 83.8 % for the RBF-SVM, and 83.6 % for the HIK-SVM. The SIFT-ScSPM algorithm also has higher values of precision, recall, F-measure, and *AUC* than other algorithms.

The SIFT-LLC and SIFT-FV perform better than the HOG-SPM and the SIFT-SPM. The SIFT-LLC uses the locality-constrained linear coding and the SIFT-FV uses Fisher kernel coding, whereas the HOG-SPM and the SIFT-SPM uses vector quantization for global feature formation. The result indicates that a better encoding algorithm like ScSPM and Fisher vector improves the classification performance. Note that among the top-seven algorithms, SIFT-ScSPM, SIFT-LLC, SIFT-FV, HOG-SPM, SIFT-SPM, and SURF-ScSPM all encode local features using BoW.

It is interesting to note that the classification rates of HOG-SPM and SIFT-SPM improve by 8.3 and 12.0 %, respectively, when using the RBF-SVM compared to using the linear-SVM. In addition, using the HIK-SVM increases the classification rates of HOG-SPM and SIFT-SPM by 9.4 and 17.0 %, respectively. However, for the SIFT-LLC, SIFT-ScSPM, and SIFT-FV, the classification performance is not improved by using the RBF-SVM and HIK-SVM. Note that previous tests (Yang et al. 2009; Wang et al. 2010) on several benchmark data sets also indicate that the choice of the SVM kernel does not affect the performance of the LLC and ScSPM significantly.

The SURF-ScSPM is also a BOW-based descriptor. However, it extracts local features from the interest point patches. The *CRs* of SURF-ScSPM is lower than the *CRs* of SIFT-ScSPM, SIFT-LLC, and HOG-SPM, which extract local features from the dense patches. The result indicates that sparse local features can not cover enough information for scene categorization compared to dense local features.

Table 4. Scene categorization performance on the 15-scene data set using the linear-SVM

ID	Algorithms	CR (%)	Precision (%)	Recall (%)	F-measure (%)	AUC (%)
1	SIFT-ScSPM	**84.5 ± 1.5**	**84.8 ± 1.7**	**84.0 ± 1.4**	**84.1 ± 1.4**	**98.9 ± 0.1**
2	SIFT-LLC	83.0 ± 1.3	83.0 ± 1.5	82.2 ± 1.5	82.3 ± 1.4	98.8 ± 0.2
3	SIFT-FV	80.2 ± 1.8	79.6 ± 1.9	79.2 ± 1.9	79.0 ± 2.0	98.4 ± 0.1
4	HOG-SPM	70.6 ± 2.7	71.7 ± 2.7	68.7 ± 2.6	70.0 ± 2.4	96.3 ± 0.3
5	OB	79.9 ± 1.6	80.0 ± 1.6	79.1 ± 2.0	79.0 ± 1.9	97.7 ± 0.3
6	SIFT-SPM	60.9 ± 4.0	62.3 ± 5.6	57.8 ± 4.3	57.0 ± 4.7	94.4 ± 0.5
7	SURF-ScSPM	73.3 ± 2.1	72.9 ± 2.1	72.3 ± 2.3	72.3 ± 2.4	97.3 ± 0.4
8	GIST	71.5 ± 1.2	71.1 ± 1.4	70.6 ± 1.5	71.2 ± 1.0	95.7 ± 0.4
9	CENTRIST	72.7 ± 1.4	72.8 ± 1.1	72.2 ± 1.9	71.9 ± 1.7	95.9 ± 0.3
10	LBP	71.1 ± 2.9	69.8 ± 4.0	70.3 ± 3.4	69.3 ± 3.9	94.9 ± 1.2
11	Uniform LBP	56.2 ± 3.0	55.6 ± 5.8	54.7 ± 3.2	52.6 ± 3.6	92.8 ± 1.0
12	LBP-HF	64.9 ± 3.4	63.8 ± 4.6	63.9 ± 3.8	62.5 ± 4.3	92.8 ± 1.0
13	PLBP	53.8 ± 3.6	52.7 ± 4.7	52.4 ± 3.6	51.5 ± 4.1	92.1 ± 1.5
14	HMAX	61.1 ± 4.1	60.8 ± 4.0	59.8 ± 4.5	59.9 ± 4.3	79.7 ± 6.9

Table 5. Scene categorization performance on the 15-scene data set using the RBF-SVM

ID	Algorithms	CR (%)	Precision (%)	Recall (%)	F-measure (%)	AUC (%)
1	SIFT-ScSPM	**83.8 ± 1.7**	**84.2 ± 1.6**	**83.2 ± 1.6**	**83.5 ± 1.3**	**98.3 ± 0.0**
2	SIFT-LLC	82.2 ± 1.1	82.5 ± 1.2	81.5 ± 1.1	81.5 ± 1.2	98.2 ± 0.1
3	SIFT-FV	79.4 ± 1.7	78.9 ± 2.1	78.1 ± 2.1	77.6 ± 2.3	97.5 ± 0.2
4	HOG-SPM	78.9 ± 1.2	76.8 ± 3.2	76.3 ± 2.9	76.1 ± 3.1	96.5 ± 0.5
5	OB	73.2 ± 2.0	73.5 ± 1.9	72.0 ± 2.4	72.1 ± 2.3	95.5 ± 0.4
6	SIFT-SPM	72.9 ± 1.3	72.0 ± 1.6	71.6 ± 1.4	71.3 ± 1.4	95.2 ± 0.4
7	SURF-ScSPM	72.4 ± 1.6	71.6 ± 1.8	71.1 ± 1.7	70.6 ± 1.8	95.8 ± 0.4
8	GIST	72.8 ± 1.2	72.2 ± 1.0	71.8 ± 1.4	71.5 ± 1.2	95.2 ± 0.2
9	CENTRIST	72.6 ± 1.8	72.7 ± 1.1	72.0 ± 2.1	71.8 ± 1.8	91.2 ± 1.0
10	LBP	70.9 ± 4.2	69.6 ± 5.2	70.1 ± 4.8	69.2 ± 5.3	96.2 ± 0.7
11	Uniform LBP	67.8 ± 3.9	66.8 ± 5.0	66.8 ± 4.4	65.6 ± 4.9	94.0 ± 0.8
12	LBP-HF	67.3 ± 4.4	66.0 ± 5.4	66.5 ± 4.9	66.3 ± 4.4	94.0 ± 0.8
13	PLBP	69.4 ± 3.5	69.2 ± 4.9	68.2 ± 4.1	68.1 ± 4.4	95.4 ± 1.0
14	HMAX	62.4 ± 3.7	61.6 ± 3.4	61.0 ± 4.0	60.8 ± 3.7	79.5 ± 6.0

Table 6. Scene categorization performance on the 15-scene data set using the HIK-SVM

ID	Algorithms	CR (%)	Precision (%)	Recall (%)	F-measure (%)	AUC (%)
1	SIFT-ScSPM	**83.6 ± 1.6**	**84.2 ± 1.7**	**83.2 ± 1.4**	**83.2 ± 1.4**	**98.2 ± 0.1**
2	SIFT-LLC	82.6 ± 1.5	82.8 ± 1.4	82.1 ± 1.3	82.1 ± 1.3	98.2 ± 0.0
3	SIFT-FV	80.0 ± 1.7	80.1 ± 2.0	79.7 ± 1.4	79.3 ± 1.7	97.7 ± 0.2
4	HOG-SPM	80.0 ± 2.6	79.5 ± 2.5	79.7 ± 2.5	79.6 ± 2.4	97.8 ± 0.2
5	OB	76.9 ± 2.0	76.9 ± 1.8	76.0 ± 2.3	76.0 ± 2.2	97.4 ± 0.2
6	SIFT-SPM	77.9 ± 1.2	77.5 ± 1.3	76.8 ± 1.0	76.8 ± 1.0	97.0 ± 0.3
7	SURF-ScSPM	72.3 ± 2.4	71.4 ± 2.5	71.2 ± 2.5	70.6 ± 2.6	95.6 ± 0.4
8	GIST	72.1 ± 0.7	71.8 ± 0.7	71.4 ± 0.9	72.1 ± 0.9	95.2 ± 0.2
9	CENTRIST	70.9 ± 2.4	71.3 ± 1.8	70.1 ± 2.6	70.0 ± 2.3	95.9 ± 0.4
10	LBP	71.9 ± 2.5	71.0 ± 3.2	71.1 ± 3.2	70.5 ± 3.4	95.7 ± 0.8
11	Uniform LBP	70.6 ± 3.5	70.4 ± 4.4	69.8 ± 4.0	69.3 ± 4.5	95.7 ± 1.0
12	LBP-HF	66.4 ± 3.3	65.6 ± 4.4	65.5 ± 3.7	64.8 ± 4.2	95.7 ± 1.0
13	PLBP	73.0 ± 3.4	72.3 ± 5.1	72.9 ± 4.2	72.6 ± 4.7	96.6 ± 1.0
14	HMAX	63.9 ± 3.4	63.5 ± 3.6	62.5 ± 3.7	62.6 ± 3.6	82.9 ± 8.3

The LBP, uniform LBP, LBP-HF, PLBP, and CENTRIST have higher classification rates than the HMAX on the 15-scene data set. The PLBP has the highest classification rate (73.0 %) among the LBP-based descriptors. For the original LBP, uniform LBP, and PLBP, scene categorization performance is better using the HIK-SVM than the linear-SVM and the RBF-SVM. The highest classification rates of LBP, uniform LBP, PLBP are 71.9, 70.6, and 73.0 %, respectively. For the LBP-HF algorithm, a higher classification rate (67.3 %) is achieved using the RBF-SVM, compared to the linear-SVM (64.9 %) and HIK-SVM (66.4 %). The highest *CR* of the CENTRIST is 72.7 %, achieved with the linear-SVM. In fact, the CENTRIST is also a LBP-based algorithm because it extracts local features using the LBP feature map. The difference is that the CENTRIST forms the global feature vector using the spatial PCA, whereas the LBP, uniform LBP, and LBP-HF form the global feature vector using histograms. The PCA with 3 spatial levels accounts for the higher performance of the CENTRIST over the original LBP algorithm.

The OB has a higher classification rate (79.9 %) than the SIFT-SPM, the LBP-based algorithms, and the biologically-inspired algorithms on the 15-scene data set. The OB algorithm achieves its highest classification rate of 79.9 % when using the linear-SVM. A similar observation was reported in (Li et al.2010); the OB performs better than the SIFT-SPM on the UIUC-sport-event data set, the 15-scene data set, and the 67-indoor scene data set. However, in our experiment, the OB has a lower *CR* than the SIFT-ScSPM and SIFT-LLC on the 15-scene data set. This result indicates that the global feature formation used in the OB is not as good as in the SIFT-ScSPM and SIFT-LLC.

Based on this experiment, we determined a suitable SVM kernel for each of the descriptors. The selected SVM classifiers were used in the subsequent experiments, where we evaluated the visual descriptors on three other data sets: the 8-outdoor-scene, the 67-indoor-scene, and the SUN397 data sets. The aim of the subsequent experiments is to identify the algorithms that had consistent performance on multiple data sets.

Table 7. Scene categorization performance on the 8-nature-outdoor-scene data set

ID	Algorithms	CR (%)	Precision (%)	Recall (%)	F-measure (%)	AUC (%)
1	SIFT-ScSPM (linear)	**89.8 ± 1.8**	**90.4 ± 1.8**	**90.0 ± 2.2**	**90.0 ± 2.1**	**98.4 ± 0.4**
2	SIFT-LLC (linear)	88.1 ± 2.1	88.6 ± 2.0	88.3 ± 2.6	88.3 ± 2.4	98.2 ± 0.6
3	SIFT-FV (linear)	88.1 ± 3.3	88.2 ± 3.4	88.0 ± 3.9	87.9 ± 3.8	98.7 ± 0.4
4	HOG-SPM(HIK)	88.1 ± 2.5	89.0 ± 2.1	88.2 ± 3.2	88.3 ± 2.9	98.3 ± 0.6
5	OB (linear)	87.5 ± 2.5	88.0 ± 2.3	87.6 ± 3.1	87.5 ± 2.8	98.4 ± 0.5
6	SIFT-SPM(HIK)	87.4 ± 2.3	87.9 ± 2.2	87.8 ± 2.5	87.6 ± 2.4	98.1 ± 0.4
7	SURF-ScSPM (linear)	85.9 ± 2.6	85.4 ± 2.8	85.0 ± 3.4	85.9 ± 3.2	98.0 ± 0.5
8	GIST (RBF)	85.3 ± 2.5	85.9 ± 2.4	85.4 ± 3.1	85.4 ± 2.9	98.0 ± 0.8
9	CENTRIST (linear)	83.5 ± 4.9	84.3 ± 4.8	83.6 ± 5.8	83.5 ± 5.5	97.2 ± 1.1
10	LBP (HIK)	76.6 ± 4.3	77.0 ± 4.6	76.8 ± 5.1	76.5 ± 4.9	95.8 ± 1.2
11	Uniform LBP (HIK)	75.2 ± 4.6	76.0 ± 4.6	75.2 ± 5.8	74.9 ± 5.5	95.0 ± 1.6
12	LBP-HF (RBF)	71.9 ± 3.7	72.1 ± 4.5	71.9 ± 4.7	71.4 ± 4.6	92.5 ± 2.6
13	PLBP (HIK)	81.4 ± 2.9	81.6 ± 3.6	81.3 ± 3.7	81.2 ± 3.8	96.7 ± 1.1
14	HMAX (HIK)	79.8 ± 0.8	80.1 ± 0.7	80.2 ± 0.7	80.0 ± 0.7	96.5 ± 0.3

4.4.2 Classification results on the 8-outdoor-scene data set

Table 7 shows scene categorization results of the 14 selected descriptors on the 8-outdoor-scene data set. Among the 14 descriptors, the SIFT-ScSPM has the highest *CR* (89.8 %). The SIFT-ScSPM, SIFT-LLC, SIFT-FV, HOG-SPM, OB, SIFT-SPM, and SURF-ScSPM are the top-7 algorithms, listed in a descending order of *CR*. For the last 7 descriptors, the classification rates of GIST, CENTRIST and PLBP are higher than 80.0 %. Note that the HMAX achieves a higher *CR* (79.8 %) on the outdoor scene categorization than three LBP-based algorithms (76.6 % for LBP, 75.2 % for uniform LBP, and 71.9 % for LBP-HF). This result indicates that the biologically-inspired features are useful for outdoor scene categorization.

4.4.3 Classification results on the 67-indoor-scene data set

Table 8 shows scene categorization results of the 14 selected descriptors on the 67-indoor-scene data set. The SIFT-ScSPM still has the highest *CR* (45.6 %) compared with the other 13 descriptors. The top-7 descriptors are the SIFT-ScSPM, OB, SIFT-LLC, SIFT-FV, SURF-ScSPM, HOG-SPM, and SIFT-SPM, listed in a descending order of *CR*. These seven algorithms (except for the OB) use BoW methods for global feature formation. These results indicate that using BoW methods is more robust than using histograms, PCA and down-sampling, especially when the complexity of images is increased. Note that the OB descriptor outperformed most of the BoW methods on the 67-indoor-scene data set. This result indicates that the object information is useful for indoor scene categorization.

From Tables 4, 5, 6, 7 and 8, we can see that the ranking based on different measures were consistent for the top-seven algorithms. A higher *CR* was also accompanied by a higher precision, recall, F-measure, and*AUC* values.

4.4.4 Classification results on the SUN397 data set

Using the SUN397 data set, we compared the 14 visual descriptors and 4 recent methods based on deep learning: OverFeat (Sermanet et al. 2014), DeCAF (Donahue et al. 2014), ImageNet-CNN (Krizhevsky et al.2012), and Places-CNN (Zhou et al. 2014). Note that training a CNN for scene categorization on a large data set requires significant engineering efforts for parameter tuning. To achieve a fair comparison, we evaluated the 14 visual descriptors using the same evaluation protocol described in (Xiao et al. 2014a) for the SUN397 data set. This data set is divided into fixed partitions. In each partition, 50 training images and 50 test images per class are used for evaluation. The classification rate, averaged over the fixed partitions, is used for comparison. The four deep learning methods had been evaluated using the same protocol, and their results have been reported in (Donahue et al. 2014; Zhou et al. 2014; Xiao et al. 2014a).

Figure 9 presents the *CRs* and their standard deviation of the 14 visual descriptors and the four deep-learning methods (OverFeat, DeCAF, ImageNet-CNN, and Places-CNN) on the SUN397 data sets. All the four deep features outperform the other visual descriptors. However, even the best algorithm (Places-CNN) had a *CR* of only 54.3 %, which is still significantly lower than human performance of 68.0 % (Xiao et al.2014a).

To rank the data sets in terms of their degree of difficulty, we compared the average classification rates of the top 7 descriptors on each of the four data set. The highest average *CR* of 87.8 % is obtained with the 8-outdoor-scene data set, compared to 79.8 % for the 15-scene data set, 39.0 % for the 67-indoor-scene data set, and 25.4 % for the SUN397 data set. These results indicate that among the four data sets, the 8-outdoor-scene data set is the easiest and the SUN397 data set is the hardest to classify. Similar ranking is obtained if we use the median *CR* on each data set. Apart from the difference in the number of images and scene categories, the 8-outdoor-scene data set consists of only outdoor images, whereas the SUN397 data set contains not only outdoor scenes but also indoor and man-made scenes.

4.5 Class separability and stability of feature vectors

We evaluated the class separability of the feature vectors using the Fisher score *S* (see Sect. 3.2). Note that this evaluation is independent of the classifier used.

Table 8. Scene categorization performance on the 67-indoor data set

ID	Algorithms	CR (%)	Precision (%)	Recall (%)	F-measure (%)	AUC (%)
1	SIFT-ScSPM (linear)	**45.6 ± 1.0**	**45.4 ± 2.5**	**35.8 ± 1.5**	**36.9 ± 1.7**	**91.7 ± 0.4**
2	SIFT-LLC (linear)	43.9 ± 0.4	44.7 ± 1.2	35.2 ± 0.8	36.3 ± 0.9	91.3 ± 0.5
3	SIFT-FV (linear)	41.5 ± 1.2	46.0 ± 1.3	31.4 ± 1.0	32.5 ± 1.0	90.4 ± 0.6
4	HOG-SPM (HIK)	31.0 ± 0.5	27.5 ± 0.8	25.5 ± 0.6	25.8 ± 0.6	85.1 ± 0.6
5	OB (linear)	45.3 ± 0.5	43.7 ± 1.0	39.4 ± 0.7	39.2 ± 0.7	91.7 ± 0.4
6	SIFT-SPM (HIK)	31.4 ± 0.8	28.4 ± 1.1	25.9 ± 0.7	26.4 ± 0.8	85.3 ± 0.5
7	SURF-ScSPM (linear)	34.5 ± 0.7	35.5 ± 0.9	25.4 ± 0.4	26.2 ± 0.2	87.6 ± 0.3
8	GIST (RBF)	30.9 ± 0.9	28.0 ± 0.1	25.2 ± 0.3	25.7 ± 0.4	84.4 ± 0.4
9	CENTRIST (linear)	12.2 ± 11.0	15.4 ± 8.5	10.8 ± 9.6	10.1 ± 9.6	78.4 ± 3.8
10	LBP (HIK)	22.9 ± 0.6	21.0 ± 1.0	17.7 ± 0.9	18.4 ± 0.9	81.7 ± 0.5
11	Uniform LBP (HIK)	22.0 ± 1.1	18.5 ± 0.4	16.2 ± 1.0	16.6 ± 0.9	80.0 ± 0.6
12	LBP-HF (RBF)	15.4 ± 1.2	12.7 ± 0.8	11.9 ± 0.7	11.8 ± 0.7	76.3 ± 1.0
13	PLBP (HIK)	27.2 ± 1.0	24.1 ± 1.3	21.6 ± 1.0	22.2 ± 1.0	84.5 ± 0.3
14	HMAX (HIK)	11.6 ± 2.3	10.5 ± 2.7	9.4 ± 1.9	9.5 ± 2.1	72.9 ± 3.0

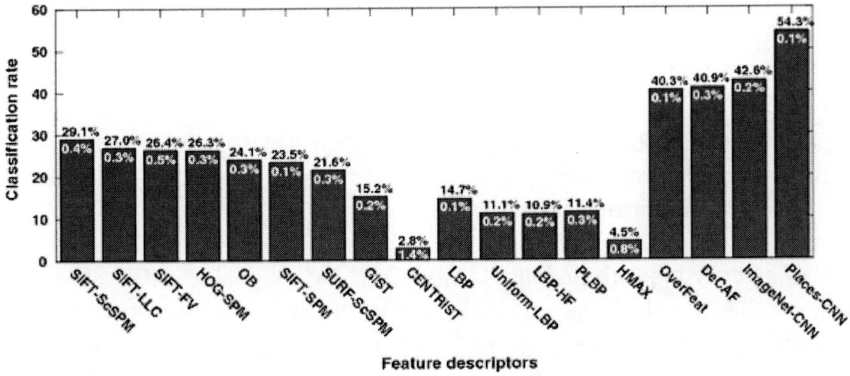

Figure 9. Comparison of scene categorization methods on the SUN397 data set. For each method, the *top number* in*black* is the classification rate, and the *bottom number* in *white* is the standard deviation.

Table 9 presents the class separability scores (*S*) for the compared features on the four data sets. Among the biologically-inspired features, the HMAX has a low *S* score on the four data sets. The HMAX features are formed by using max pooling in layer C1, BoW in layer S2, and max pooling in layer C2. The GIST has a higher *S* score than the HMAX, CENTRIST, OB, and the LBP-based features. As shown in Tables 4, 5, 6, 7and 8, the GIST algorithm also has higher classification rates than the OB, HMAX, LBP, uniform LBP, and LBP-HF. However, the GIST algorithm has a lower class separability score than BoW algorithms (SIFT-ScSPM, SURF-ScSPM, SIFT-LLC, SIFT-SPM, and HOG-SPM). Note that the GIST forms global features using only down-sampling and averaging. This result indicates that biologically-inspired features can benefit from better schemes for forming global features.

Among the 14 descriptors, the SIFT-FV (which calculates SIFT features on dense-grids, and forms global features using the Fisher kernel coding) has the highest *S* score on the 8-outdoor-scene data set (2.5470), the 15-outdoor-scene data set (2.3111), the 67-indoor-scene data set (1.1484), and the SUN397 data set (1.0132). The result indicates the Fisher vector extracts discriminative global features. Note that in our experiment, the SIFT-FV has lower *CRs* than the SIFT-ScSPM on the four data sets. However, in the paper of Sanchez et al. (2013), the *CR* of SIFT-FV on the SUN397 data set is 43.3 %, which is higher than other hand-designed features compared in our experiments. The reason is the SIFT-FV in (Sanchez et al. 2013) extracts local SIFT features from overlapping patches (24×2424×24) on a regular grid every 4 pixels at 5 scales. The SIFT-FV in our experiment extracts SIFT features from overlapping patches (16×1616×16) on a regular grid every 8 pixels at one scale. The difference between the *CRs* of SIFT-FV in (Sanchez et al. 2013) and in our paper indicates that informative local features and discriminative global feature formation methods can improve the classification performance.

The other BoW algorithms (SIFT-LLC, SIFT-ScSPM, SIFT-SPM, HOG-SPM, and SURF-ScSPM) also has higher *S* scores than the LBP-based algorithms (LBP, uniform LBP, LBP-HF, PLBP, and CENTRIST) on the four data sets. Note that to form global features, the SIFT-ScSPM and SIFT-LLC combine the BoW algorithms with spatial histograms and max pooling, whereas the LBP-based methods only use histograms.

The SURF-ScSPM has lower *S* score than most of the BOW-based descriptors, such as SIFT-ScSPM, SIFT-LLC, and SIFT-FV, but it has higher *S* score than OB, SIFT-SPM, LBP-based, and biologically-inspired descriptors. This shows that compared to the SIFT-ScSPM, the discriminative power of the SURF-ScSPM is reduced by extracting the SURF features from key points. The *S* score of SURF-ScSPM is 1.8323 on the 8-outdoor-scene data set, 1.7330 on the 15-outdoor-scene data set, 1.0525 on the 67-indoor-scene data set, and 1.0060 on the SUN397 data set.

The LBP-based features (LBP, uniform LBP, PLBP, LBP-HF, and CENTRIST) have lower *S* scores than the BoW features. However, the LBP-based features achieve higher *S* scores than the OB and HMAX. Note that the LBP-based features are more efficient to compute than the BoW features. Among the LBP-based algorithms, the PLBP achieved the highest class separability score. This is also reflected in the higher *CR* of the PLBP, compared to the LBP, uniform LPB, LBP-HF, and CENTRIST (see Sect. 4.4). The uniform LBP and the LBP-HF reduce the dimension and also the class separability of features. This result indicates that the class separability and classification accuracy of the LBP-based algorithms can be improved by using a better global feature formation, instead of the histograms.

The OB has the lowest *S* score among the compared features. Its *S* score is 1.0003 on the 8-outdoor-scene data set, 1.0001 on the 15-outdoor-scene data set, 1.0001 on the 67-indoor-scene data set, and 1.0000 on the SUN397 data set. The reason may be that many similar objects appear in different scene categories. Note that the OB uses only max pooling to form global features, and it extracts a large number of features (44,604 per image).

Table 9. The *S* score for class separability of feature vectors

ID	Algorithms	8-outdoor-scene	15-scene	67-indoor-scene	SUN397
1	SIFT-ScSPM	1.8827	1.3005	1.0364	1.0064
2	SIFT-LLC	2.4339	1.5626	1.1441	1.0061
3	SIFT-FV	**2.5470**	**2.3111**	**1.1484**	**1.0132**
4	HOG-SPM	1.6340	1.6010	1.0959	1.0042
5	OB	1.0003	1.0001	1.0001	1.0000
6	SIFT-SPM	1.4521	1.3573	1.0516	1.0052
7	SURF-ScSPM	1.8323	1.7330	1.0525	1.0060
8	CENTRIST	1.0045	1.0039	1.0000	1.0000
9	GIST	1.0604	1.0283	1.0047	1.0008
10	LBP	1.0434	1.0136	1.0021	1.0004
11	Uniform LBP	1.0187	1.0070	1.0015	1.0003
12	LBP-HF	1.0146	1.0091	1.0012	1.0002
13	PLBP	1.0630	1.0255	1.0034	1.0002
14	HMAX	1.0006	1.0003	1.0004	1.0001

A high value of *S* means the extracted scene categories are highly separable using the given feature vector

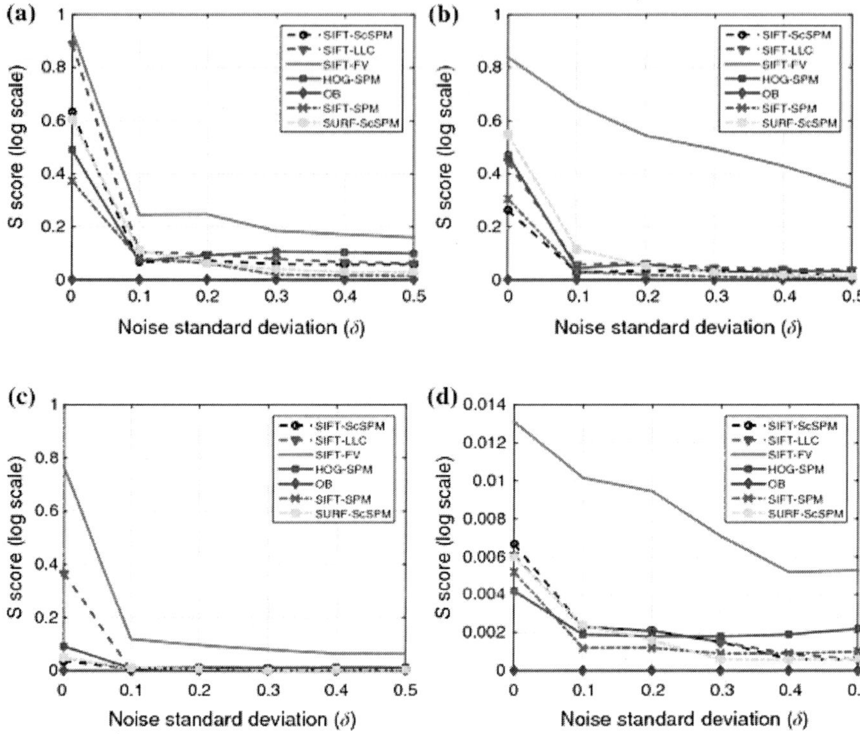

Figure 10. Stability of features under the presence of Gaussian noise of varying standard deviation, on the four data sets. **a** 8-scene data set, **b** 15-scene data set, **c** 67-indoor-scene data set, **d** SUN397 data set

Next, we evaluated the stability of the scene categorization algorithms in the presence of noise. In this experiment, Gaussian noise with varied standard deviation was added to the original images. Then, the class separability scores were computed for the noisy images. Figure 10 shows the results for the top-7 algorithms: SIFT-ScSPM, SIFT-LLC, SIFT-FV, HOG-SPM, OB, SIFT-SPM, and SURF-ScSPM. These algorithms (based on the bag-of-words) are identified as having high classification accuracy in Sect. 4.4. When noise is added, the class separability (*S* score) of all features reduces. At all noise levels, the SIFT-FV descriptors has a higher *S* score than all other descriptors.

The results presented in this section indicate that the method for forming global features affects the class separability significantly. Using BoW algorithms before applying histograms, PCA or max pooling (as in SIFT-FV,

SIFT-ScSPM, and HOG-SPM) produces feature vectors with more discriminative power. Using histograms as the first step of the global feature formation (as in CENTRIST) decreases the class separability of features. Using only one method for global feature formation (as in the LBP-based algorithms) does not yield high separability scores and nor high classification rates.

5. CONCLUSION

This paper presented an experimental evaluation of existing visual descriptors for scene categorization. Fourteen descriptors were grouped into three categories: biologically-inspired features, local features, and global feature formation. The existing benchmark data sets and performance measures for scene categorization were also discussed.

The experimental results indicate that the SIFT-ScSPM outperforms all other tested descriptors on the 15-scene data set. The SIFT-ScSPM uses SIFT as its local descriptor and ScSPM as its global feature formation. Local descriptors, SIFT, HOG, and SURF achieve higher classification rates than LBP, CENTRIST, and HMAX. The global feature formation methods affect the class separability of feature vectors significantly. Using ScSPM, LLC, and FV for global feature formation leads to higher class separability than using histograms and PCA. Using BoW before histograms, PCA, and max pooling makes feature vectors more distinguishable. The results on 67-indoor-scene data set show that the mid-level features like objects, bag-of-parts, and the efficient patch encoding algorithm like Fisher vector improve the classification rates for indoor scenes. The results on the SUN397 data set indicate that the SIFT-ScSPM outperforms all other hand-designed descriptors. The learned features produced by deep learning establish the new state-of-the art performance in scene categorization. However, there is still a large performance gap between the best computational algorithm and humans.

Based on this survey and evaluation, several promising research directions can be highlighted. First, local feature descriptors can be built that combine the properties of SIFT, HOG, SURF, GIST, or the early stages of deep learning architecture. A good local descriptor leads to a high classification rate. Second, global feature formation algorithms can be developed based on ScSPM and FV. Third, most existing studies on gist recognition have been concerned with static scenes, which is the focus of this study. In recent years, gist recognition of dynamic scenes has attracted the attention of researchers (Shroff et al. 2010; Derpanis et al. 2012; De Geest and Tuytelaars 2004; Ng et al. 2015), and therefore, the extension of this study to dynamic scenes would be invaluable.

REFERENCES

1. Abe Y, Shikano M, Fukuda T, Arai F, Tanaka Y (1999) Vision based navigation system for autonomous mobile robot with global matching. In: Proceedings of IEEE international conference on robotics and automation, pp 1299–1304

2. Ahonen T, Hadid A, Pietikainen M (2006) Face description with local binary patterns: application to face recognition. IEEE Trans Pattern Anal Mach Intell 28(12):2037–2041

3. Ahonen T, Matas J, He C, Pietikainen M (2009) Rotation invariant image description with local binary pattern histogram Fourier features. In: Proceedings of Scandinavian conference on image analysis, pp 61–70

4. Alahi A, Ortiz R, Vandergheynst P (2012) FREAK: fast retina keypoint. In: Proceedings of IEEE conference on computer vision and pattern recognition, pp 510–517

5. Bai Y, Guo L, Jin L, Huang Q (2009) A novel feature extraction method using pyramid histogram of orientation gradients for smile recognition. In: Proceedings of IEEE international conference on image processing, pp 3305–3308

6. Bao SY, Sun M, Savarese S (2011) Toward coherent object detection and scene layout understanding. Image Vis Comput 29(9):569–579

7. Bay H, Ess A, Tuytelaars T, Van Gool L (2008) Speeded-up robust features (SURF). Comput Vis Image Underst 110(3):346–359

8. Bergamo A, Torresani L (2014) Classemes and other classifier-based features for efficient object categorization. IEEE Trans Pattern Anal Mach Intell 36(10):1988–2001

9. Blei D, Ng A, Jordan M (2003) Latent Dirichlet allocation. J Mach Learn Res 3:993–1022

10. Bo L, Ren X, Fox D (2010) Kernel descriptors for visual recognition. In: Lafferty J, Williams C, Shawe-Taylor J, Zemel RS, Culotta A (eds) Advances in neural information processing systems 23, Curran Associates, Inc., pp 244–252

11. Boiman O, Shechtman E, Irani M (2008) In defense of nearest-neighbor based image classification. In: Proceedings of IEEE conference on computer vision and pattern recognition, pp 1–8

12. Boutell MR, Jiebo L, Brown CM (2006) Factor graphs for region-based whole-scene classification. In: Proceedings of IEEE conference on computer vision and pattern recognition workshop, pp 104–104

13. Brown M, Susstrunk S (2011) Multi-spectral SIFT for scene category recognition. In: Proceedings of IEEE conference on computer vision and pattern recognition, pp 177–184

14. Brumby SP, Kenyon G, Landecker W, Rasmussen C, Swaminarayan S, Bettencourt LMA (2009) Large-scale functional models of visual cortex for remote sensing. In: Proceedings of IEEE applied imagery pattern recognition workshop, pp 1–6

15. Burges CJC (1998) A tutorial on support vector machines for pattern recognition. Data Min Knowl Disc 2(2):121–167

16. Cakir F, Gudukbay U, Ulusoy O (2011) Nearest-neighbor based metric functions for indoor scene recognition. Comput Vis Image Underst 115(11):1483–1492

17. Castelhano MS, Henderson JM (2008) The influence of color on the perception of scene gist. J Exp Psychol Hum Percept Perform 34(3):660–675

18. Chang CK, Siagian C, Itti L (2010) Mobile robot vision navigation and localization using gist and saliency. In: Proceedings of IEEE-RSJ international conference on intelligent robots and systems, pp 4147–4154

19. Chella A, Frixione M, Gaglio S (2001) Conceptual spaces for computer vision representations. Artif Intell Rev 16(2):137–152

20. Cheng G, Han J, Guo L, Liu T (2015) Learning coarse-to-fine sparselets for efficient object detection and scene classification. In: Proceedings of IEEE conference on computer vision and pattern recognition, pp 1173–1181

21. Chin T, Suter D, Wang H (2011) Boosting histograms of descriptor distances for scalable multiclass specific scene recognition. Image Vis Comput 29(4):241–250

22. Clinchant S, Csurka G, Perronnin F, Renders J (2007) XRCEs participation to ImagEval. In: Proceedings of workshop at content visualization and intermedia representations, pp 1–8

23. Dai Q, Zhao RW, Wu Z, Wang X, Gu Z, Wu W, Jiang YG (2015) Detecting violent scenes and affective impact in movies with deep learning. In: Proceedings of MediaEval workshop, pp 1–3

24. Dalal N, Triggs B (2005) Histograms of oriented gradients for human detection. In: Proceedings of IEEE conference on computer vision and pattern recognition, pp 886–893

25. De Geest R, Tuytelaars T (2014) Dense interest features for video processing. In: Proceedings of IEEE international conference on image processing, pp 5771–5775

26. Derpanis KG, Lecce M, Daniilidis K, Wildes RP (2012) Dynamic scene understanding: the role of orientation features in space and time in scene classification. In: Proceedings of IEEE international conference on computer vision and pattern recognition, pp 1306–1313

27. Dixit M, Chen S, Gao D, Rasiwasia N, Vasconcelos N (2015) Scene classification with semantic fisher vectors. In: Proceedings of IEEE conference on computer vision and pattern recognition, pp 2974–2983

28. Doersch C, Gupta A, Efros AA (2013) Mid-level visual element discovery as discriminative mode seeking. In: Burges C, Bottou L, Welling M, Ghahramani Z, Weinberger K (eds) Advances in neural information processing systems 26, Curran Associates, Inc., pp 494–502

29. Donahue J, Jia Y, Vinyals O, Hoffman J, Zhang N, Tzeng E, Darrell T (2014) DeCAF: a deep convolutional activation feature for generic visual recognition. In: Proceedings of international conference on machine learning, pp 647–655

30. Douze M, Jegou H, Sandhawalia H, Amsaleg L, Schmid C (2009) Evaluation of gist descriptors for web-scale image search. In: Proceedings of international conference on image and video retrieval, pp 140–147

31. Ebied HM (2012) Feature extraction using PCA and kernel-PCA for face recognition. In: Proceedings of international conference on informatics and systems, pp 72–77

32. Elfiky NM, Khan SF, Weijer J, Gonzalez J (2012) Discriminative compact pyramids for object and scene recognition. Pattern Recognit 45(4):1627–1636

33. Fawcett T (2006) An introduction to ROC analysis. Pattern Recognit Lett 27(8):861–874

34. Fei-Fei L, Perona P (2005) A Bayesian hierarchical model for learning natural scene categories. In: Proceedings of IEEE conference on computer vision and pattern recognition, pp 524–531

35. Fei-Fei L, Fergus R, Perona P (2007a) Learning generative visual models from few training examples: an incremental Bayesian approach tested on 101 object categories. J Comput Vis Image Underst 106(1):59–70

36. Fei-Fei L, Lyer A, Koch C, Perona P (2007b) What do we perceive in a glance of a real-world scene? J Vis 7(1):1–29

37. Felzenszwalb P, Girshick R, McAllester D, Ramanan D (2010) Object detection with discriminatively trained part-based models. IEEE Trans Pattern Anal Mach Intell 32(9):1627–1645

38. Fuentes-Pacheco J, Ruiz-Ascencio J, Rendon-Mancha JM (2015) Visual simultaneous localization and mapping: a survey. Artif Intell Rev 43(1):55–81

39. Funt BV, Finlayson GD (1995) Color constant color indexing. IEEE Trans Pattern Anal Mach Intell 17(5):522–529

40. Gemert JCV, Geusebroek JM, Veenman CJ, Smeulders AWM (2008) Kernel codebooks for scene categorization. In: Proceedings of European conference on computer vision, pp 696–709

41. Goh H, Thome N, Cord M, Lim JH (2014) Learning deep hierarchical visual feature coding. IEEE Trans Neural Netw Learn Syst 25(12):2212–2225

42. Gokalp D, Aksoy S (2007) Scene classification using bag-of-regions representations. In: Proceedings of IEEE conference on computer vision and pattern recognition, pp 1–8

43. Goldstein EB (2010) Encyclopedia of perception. SAGE, Thousand Oaks

44. Gowsikhaa D, Abirami S, Baskaran R (2014) Automated human behavior analysis from surveillance videos: a survey. Artif Intell Rev 42(4):747–765

45. Grossberg S, Huang T (2009) ARTSCENE: a neural system for natural scene classification. J Vis 9(4):1–19

46. Guangda S, Cuiping Z, Rong D, Cheng D (2002) MMP-PCA face recognition method. Electron Lett 38(25):1654–1656

47. Guo Z, Zhang D, Zhang D (2010) A completed modeling of local binary pattern operator for texture classification. IEEE Trans Image Process 19(6):1657–1663

48. Gupta P, Arrabolu SS, Brown M, Savarese S (2009) Video scene categorization by 3D hierarchical histogram matching. In: Proceedings of IEEE international conference on computer vision, pp 1655–1662

49. Hadjidemetriou E, Grossberg MD, Nayar SK (2004) Multiresolution histograms and their use for recognition. IEEE Trans Pattern Anal Mach Intell 26(7):831–847

50. Han X, Chen Y (2009) Image categorization by learned PCA subspace of combined visual-words and low-level features. In: Proceedings of international conference on intelligent information hiding and multimedia signal processing, pp 1282–1285

51. Han Y, Liu G (2010) A hierarchical GIST model embedding multiple biological feasibilities for scene classification. In: Proceedings of international conference on pattern recognition, pp 3109–3112

52. Hinton GE, Osindero S, Teh YW (2006) A fast learning algorithm for deep belief nets. Neural Comput 18(7):1527–1554

53. Hoiem D, Efros AA, Hebert M (2006) Putting objects in perspective. In: Proceedings of IEEE conference on computer vision and pattern recognition, pp 2137–2144

54. Huang D, Zhu C, Wang Y, Chen L (2014a) HSOG: a novel local image descriptor based on histograms of the second-order gradients. IEEE Trans Image Process 23(11):4680–4695

55. Huang Y, Wu Z, Wang L, Tan T (2014b) Feature coding in image classification: a comprehensive study. IEEE Trans Pattern Anal Mach Intell 36(3):493–506

56. Huang Z, Huang H, Zhang W, Hou L (2008) Face recognition using the global image features based on scene gist. J Inf Comput Sci 5(2):919–928

57. Hubel DH, Wiesel TN (1968) Receptive fields and functional architecture of monkey striate cortex. J Physiol 1(195):215–243

58. Igarashi M, Mizuno A, Ikebe M (2013) Accuracy improvement of histogram-based image filtering. In: IEEE international conference on image processing, pp 1217–1221

59. INRIA Graffiti data set (2004) Viewpoint change sequences. http://kahlan.eps.surrey.ac.uk/featurespace/web/data.htm

60. Jeong S, Won CS, Gray RM (2003) Histogram-based image retrieval using Gauss mixture vector quantization. In: Proceedings of IEEE international conference on acoustics, speech, and signal processing, pp 677–680

61. Jia D, Wei D, Socher R, Li-Jia L, Kai L, Fei-Fei L (2009) ImageNet: A large-scale hierarchical image database. In: Proceedings of IEEE conference on computer vision and pattern recognition, pp 248–255

62. Jia X, Schwing AG, Urtasun R (2014) Tell me what you see and i will show you where it is. In: Proceedings of IEEE conference on computer vision and pattern recognition, pp 3190–3197

63. Jiang A, Wang C, Xiao B, Dai R (2010) A new biologically inspired feature for scene image classification. In: Proceedings of international conference on pattern recognition, pp 758–761

64. Juneja M, Vedaldi A, Jawahar C, Zisserman A (2013) Blocks that shout: distinctive parts for scene classification. In: Proceedings of IEEE conference on computer vision and pattern recognition, pp 923–930

65. Kadir T, Brady M (2001) Scale, saliency and image description. Int J Comput Vis 45(2):83–105

66. Karayev S, Fritz M, Darrell T (2014) Anytime recognition of objects and scenes. In: Proceedings of IEEE conference on computer vision and pattern recognition, pp 572–579

67. Kass M, Solomon J (2010) Smoothed local histogram filters. ACM Trans Graph 29(4):100:1–100:10

68. Kazuhiro H (2011) Local autocorrelation of similarities with subspaces for shift invariant scene classification. Pattern Recogn 44(4):794–799

69. Ke Y, Sukthankar R (2004) PCA-SIFT: a more distinctive representation for local image descriptors. In: Proceedings of IEEE conference on computer vision and pattern recognition, pp 506–513

70. Khosla A, Byoungkwon A, Lim JJ, Torralba A (2014) Looking beyond the visible scene. In: Proceedings of IEEE conference on computer vision and pattern recognition, pp 3710–3717

71. Kikuchi H, Kataoka S, Muramatsu S, Huttunen H (2013) Color-tone similarity of digital images. In: Proceedings of IEEE international conference on image processing, pp 393–397

72. Koenderink JJ, Van Doorn AJ (1999) The structure of locally orderless images. Int J Comput Vis 31(2):159–168

73. Krapac J, Verbeek J, Jurie F (2011) Modeling spatial layout with fisher vectors for image categorization. In: Proceedings of IEEE international conference on computer vision, pp 1487–1494

74. Krizhevsky A, Sutskever I, Geoffrey EH (2012) ImageNet classification with deep convolutional neural networks. In: Pereira F, Burges C, Bottou L,Weinberger K (eds) Advances in neural information processing systems 25, Curran Associates, Inc., pp 1097–1105

75. Lazebnik S, Schmid C, Ponce J (2006) Beyond bags of features: spatial pyramid matching for recognizing natural scene categories. In:

Proceedings of IEEE conference on computer vision and pattern recognition, pp 2169–2178

76. LeCun Y, Bottou L, Bengio Y, Haffner P (1998) Gradient-based learning applied to document recognition. Proc IEEE 86(11):2278–2324

77. Lee HY, Lee HK, Ha YH (2003) Spatial color descriptor for image retrieval and video segmentation. IEEE Trans Multimedia 5(3):358–367

78. Lee KW, Choo H (2013) A critical review of selective attention: an interdisciplinary perspective. Artif Intell Rev 40(1):27–50

79. Lee TS, Mumford D (2003) Hierarchical Bayesian inference in the visual cortex. J Opt Soc Am 20(7):1434–1448

80. Leibe B, Leonardis A, Schiele B (2004) Combined object categorization and segmentation with an implicit shape model. In: Proceedings of European conference on computer vision workshop, pp 17–32

81. Leutenegger S, Chli M, Siegwart RY (2011) BRISK: binary robust invariant scalable keypoints. In: Proceedings of IEEE international conference on computer vision, pp 2548–2555

82. Li L, Fei-Fei L (2007) What, where and who? Classifying events by scene and object recognition. In: Proceedings of IEEE international conference on computer vision, pp 1–8

83. Li L, Su H, Lim Y, Fei-Fei L (2010) Object bank: a high-level image representation for scene classification and semantic feature sparsification. In: Lafferty J, Williams C, Shawe-Taylor J, Zemel R, Culotta A (eds) Advances in neural information processing systems 23, Curran Associates, Inc., pp 1378–1386

84. Li T, Zhu S, Ogihara M (2003) Using discriminant analysis for multi-class classification. In: Proceedings of IEEE international conference on data mining, pp 589–592

85. Li Z, Liu G, Yang Y, You J (2012) Scale- and rotation-invariant local binary pattern using scale-adaptive texton and subuniform-based circular shift. IEEE Trans Image Process 21(4):2130–2140

86. Linsley D, MacEvoy SP (2014) Evidence for participation by object-selective visual cortex in scene category judgments. J Vis 14(9):1–17

87. Liu C, Yuen J, Torralba A (2011) Sift flow: dense correspondence across scenes and its applications. IEEE Trans Pattern Anal Mach Intell 33(5):978–994

88. Liu GH, Yang JY (2013) Content-based image retrieval using color difference histogram. Pattern Recogn 46(1):188–198

89. Loschky LC, Sethi A, Simons DJ, Pydimarri TN, Ochs D, Corbeille JL (2007) The importance of information localization in scene gist recognition. J Exp Psychol Hum Percept Perform 33(6):1431–1450

90. Lowe DG (2004) Distinctive image features from scale-invariant keypoints. Int J Comput Vis 60(2):91–110

91. Maggio E, Cavallaro A (2009) Learning scene context for multiple object tracking. IEEE Trans Image Process 18(8):1873–1884

92. Maji S, Berg AC, Malik J (2008) Classification using intersection kernel support vector machines is efficient. In: Proceedings of IEEE conference on computer vision and pattern recognition, pp 1–8

93. Malcolm GL, Nuthmann A, Schyns PG (2014) Beyond gist: strategic and incremental information accumulation for scene categorization. Psychol Sci 25(5):1087–1097

94. Malhi A, Gao RX (2004) PCA-based feature selection scheme for machine defect classification. IEEE Trans Instrum Meas 53(6):1517–1525

95. Meng X, Wang Z, Wu L (2012) Building global image features for scene recognition. Pattern Recogn 45(1):373–380

96. Mikolajczyk K, Schmid C (2005) A performance evaluation of local descriptors. IEEE Trans Pattern Anal Mach Intell 27(10):1615–1630

97. Morikawa S, Shibata T (2012) Scene image recognition based on the sequence of local image vectors represented by oriented edges. In: Proceedings of IEEE international conference on acoustics, speech, and signal processing, pp 1313–1316

98. Mutch J, Lowe DG (2006) Multiclass object recognition with sparse, localized features. In: Proceedings of IEEE conference on computer vision and pattern recognition, pp 11–18

99. Ng JYH, Hausknecht M, Vijayanarasimhan S, Vinyals O, Monga R, Toderici G (2015) Beyond short snippets: deep networks for video classification. In: Proceedings of IEEE conference on computer vision and pattern recognition, pp 4694–4702

100. Ojala T, Pietikainen M, Harwood D (1996) A comparative study of texture measures with classification based on featured distributions. Pattern Recognit 29(1):51–59

101. Ojala T, Pietikainen M, Maenpaa T (2002a) Multiresolution gray-scale and rotation invariant texture classification with local binary patterns. IEEE Trans Pattern Anal Mach Intell 24(7):971–987

102. Ojala T, Pietikainen M, Maenpaa T (2002b) Multiresolution grayscale and rotation invariant texture classification with local binary patterns. IEEE Trans Pattern Anal Mach Intell 24(7):971–987

103. Oliva A (2000) Diagnostic colors mediate scene recognition. Cogn Psychol 41:176–210

104. Oliva A, Torralba A (2001) Modeling the shape of the scene: a holistic representation of the spatial envelope. Int J Comput Vis 42(3):145–175

105. Pavlopoulou C, Yu SX (2010) Indoor–outdoor classification with human accuracies: Image or edge gist? In: Proceedings of IEEE conference on computer vision and pattern recognition workshops, pp 41–47

106. Perina A, Cristani M, Murino V (2010) Learning natural scene categories by selective multi-scale feature extraction. Image Vis Comput 28(6):927–939

107. Perronnin F, Liu Y, Sanchez J, Poirier H (2010) Large-scale image retrieval with compressed fisher vectors. In: Proceedings of IEEE conference on computer vision and pattern recognition, pp 3384–3391

108. Peters RJ, Itti L (2007) Beyond bottom-up: Incorporating task-dependent influences into a computational model of spatial attention. In: Proceedings of IEEE conference on computer vision and pattern recognition, pp 1–8

109. Phillips PJ, Hyeonjoon M, Rizvi SA, Rauss PJ (2000) The FERET evaluation methodology for face-recognition algorithms. IEEE Trans Pattern Anal Mach Intell 22(10):1090–1104

110. Potter MC (1975) Meaning in visual search. Science 187(4180):965–966

111. Qian X, Hua XS, Chen P, Ke L (2011) PLBP: an effective local binary patterns texture descriptor with pyramid representation. Pattern Recognit 44(1011):2502–2515

112. Qin J, Yung NHC (2010) Scene categorization via contextual visual words. Pattern Recognit 43(5):1874–1888

113. Quattoni A, Torralba A (2009) Recognizing indoor scenes. In: IEEE conference on computer vision and pattern recognition, pp 413–420

114. Quelhas P, Monay F, Odobez JM, Gatica-Perez D, Tuytelaars T, Van Gool L (2005) Modeling scenes with local descriptors and latent aspects. In: Proceedings of IEEE international conference on computer vision, pp 883–890

115. Renninger LW, Malik J (2003) When is scene recognition just texture recognition? Vis Res 44:2301–2311

116. Riesenhuber M, Poggio T (1999) Hierarchical models of object recognition in cortex. Nat Neurosci 2:1019–1025

117. Russell BC, Torralba A, Murphy KP, Freeman WT (2008) LabelMe: a database and web-based tool for image annotation. Int J Comput Vis 77(1):157–173

118. Rybski P, Huber D, Morris D, Hoffman R (2010) Visual classification of coarse vehicle orientation using histogram of oriented gradients features. In: Proceedings of IEEE intelligent vehicles symposium, pp 921–928

119. Salzberg SL (1997) On comparing classifiers: pitfalls to avoid and a recommended approach. Data Min Knowl Disc 1(3):317–328

120. Sanchez J, Perronnin F, Mensink T, Verbeek J (2013) Image classification with the fisher vector: theory and practice. Int J Comput Vis 105(3):222–245

121. Schauerte B, Kuhn B, Kroschel K, Stiefelhagen R (2011) Multimodal saliency-based attention for object-based scene analysis. In: Proceedings of IEEE/RSJ international conference on intelligent robots and systems, pp 1173–1179

122. Scholkopf B, Sung KK, Burges CJC, Girosi F, Niyogi P, Poggio T, Vapnik V (1997) Comparing support vector machines with Gaussian kernels to radial basis function classifiers. Comput Vis Image Underst 45(11):2758–2765

123. Scholkopf B, Smola A, Muller KR (1998) Nonlinear component analysis as a kernel eigenvalue problem. Neural Comput 10(5):1299–131

124. Sermanet P, Eigen D, Zhang X, Mathieu M, Fergus R, LeCun Y (2014) OverFeat: integrated recognition, localization and detection using convolutional networks. In: Proceedings of international conference on learning representations, pp 1–15

125. Serrano N, Savakis A, Luo J (2004) Improved scene classification using efficient low-level features and semantic cues. Pattern Recognit 37(9):1773–1784

126. Serre T, Riesenhuber M (2004) Realistic modeling of simple and complex cell tuning in the hmax model, and implications for invariant object recognition in cortex. Tech. rep., Massachusetts Institute of Technology

127. Serre T, Wolf L, Poggio T (2005) Object recognition with features inspired by visual cortex. In: Proceedings of IEEE conference on computer vision and pattern recognition, pp 994–1000

128. Serre T, Wolf L, Bileschi S, Riesenhuber M, Poggio T (2007) Robust object recognition with cortex-like mechanisms. IEEE Trans Pattern Anal Mach Intell 29(3):411–426

129. Shroff N, Turaga P, Chellappa R (2010) Moving vistas: exploiting motion for describing scenes. In: Proceedings of IEEE international conference on computer vision and pattern recognition, pp 1911–1918

130. Siagian C, Itti L (2007) Biologically-inspired robotics vision Monte-Carlo localization in the outdoor environment. In: Proceedings of IEEE-

RSJ international conference on intelligent robots and systems, pp 1723–1730

131. Snowden R, Thompson P, Troscianko T (2004) Basic vision: an introduction to visual perception. Oxford University Press, New York

132. Song D, Tao D (2010) Biologically inspired feature manifold for scene classification. IEEE Trans Image Process 19(1):174–184

133. Song YJ, Park WB, Kim DW, Ahn JH (2004) Content-based image retrieval using new color histogram. In: Proceedings of international symposium on intelligent signal processing and communication systems, pp 609–611

134. Stricker M, Orengo M (1995) Similarity of color images. In: Proceedings of storage and retrieval for image and video databases, pp 381–392

135. Sun M, Huang W, Savarese S (2013) Find the best path: an efficient and accurate classifier for image hierarchies. In: Proceedings of IEEE international conference on computer vision, pp 265–272

136. Swain MJ, Ballard DH (1991) Color indexing. Int J Comput Vis 7(1):11–32

137. Theodoridis S (2015) Machine learning: a Bayesian and optimization perspective. Academic Press, London

138. Torralba A, Murphy KP, Freeman WT (2004) Sharing features: efficient boosting procedures for multiclass object detection. In: Proceedings of IEEE conference on computer vision and pattern recognition, pp 762–769

139. Torralba A, Oliva A, Castelhano M, Henderson J (2006) Contextual guidance of eye movements and attention in real-world scenes: the role of global features in object search. Psychol Rev 113(4):766–786

140. Van de Sande KEA, Gevers T, Snoek CGM (2010) Evaluating color descriptors for object and scene recognition. IEEE Trans Pattern Anal Mach Intell 32(9):1582–1596

141. Viola P, Jones M (2001) Rapid object detection using a boosted cascade of simple features. In: Proceedings of IEEE Computer Society conference on computer vision and pattern recognition, pp 511–518

142. Viola P, Jones M (2004) Robust real-time face detection. Int J Comput Vis 57(2):137–154

143. Vogel J, Schwaninger A, Wallraven C, Bülthoff HH (2007) Categorization of natural scenes: local versus global information. ACM Trans Appl Percept 4(3):19

144. Von Ahn L (2006) Games with a purpose. Computer 39(6):96–98

145. Walther DB, Chai B, Caddigan E, Beck DM, Fei-Fei L (2011) Simple line drawings suffice for functional MRI decoding of natural scene categories. Proc Nat Acad Sci 108(23):9661–9666

146. Wang J, Yang J, Yu K, Lv F, Huang T, Gong Y (2010) Locality-constrained linear coding for image classification. In: Proceedings of IEEE conference on computer vision and pattern recognition, pp 3360–3367

147. Wei L, Sang N, Wang Y (2010) A biologically inspired object-based visual attention model. Artif Intell Rev 34(2):109–119

148. Wu J, Rehg JM (2011) CENTRIST: a visual descriptor for scene categorization. IEEE Trans Pattern Anal Mach Intell 33(8):1489–1501

149. Xiao J, Ehinger KA, Hays J, Torralba A, Oliva A (2014a) SUN database: exploring a large collection of scene categories. Int J Comput Vis 1–20

150. Xiao Y, Wu J, Yuan J (2014b) mCENTRIST: a multi-channel feature generation mechanism for scene categorization. IEEE Trans Image Process 23(2):823–836

151. Xie X, Lam KM (2006) Gabor-based kernel PCA with doubly nonlinear mapping for face recognition with a single face image. IEEE Trans Image Process 15(9):2481–2492

152. Yahoo (2004) Flickr. https://www.flickr.com/

153. Yang J, Yu K, Gong Y, Huang T (2009) Linear spatial pyramid matching using sparse coding for image classification. In: Proceedings of IEEE conference on computer vision and pattern recognition, pp 1794–1801

154. Yang JL, Li HX (2010) PCA based sequential feature space learning for gene selection. In: Proceedings of international conference on machine learning and cybernetics, pp 3079–3084

155. Yao J, Fidler S, Urtasun R (2012) Describing the scene as a whole: joint object detection, scene classification and semantic segmentation. In: Proceedings of IEEE conference on computer vision and pattern recognition, pp 702–709

156. Yu K, Zhang T, Gong Y (2009a) Nonlinear learning using local coordinate coding. In: Bengio Y, Schuurmans D, Lafferty J, Williams C, Culotta A (eds) Advances in neural information processing systems 22, Curran Associates, Inc., pp 2223–2231

157. Yu Y, Mann GKI, Gosine RG (2009b) A novel robotic visual perception method using object-based attention. In: Proceedings of IEEE international conference on robotics and biomimetics, pp 1467–1473

158. Zhang H, Berg AC, Maire M, Malik J (2006) SVM-KNN: discriminative nearest neighbor classification for visual category recognition. In: Proceedings of IEEE conference on computer vision and pattern recognition, pp 2126–2136

159. Zhang L, Ji R, Xia Y, Zhang Y, Li X (2015) Learning a probabilistic topology discovering model for scene categorization. IEEE Trans Neural Netw Learn Syst 26(8):1622–1634

160. Zhang Z, Li M, Huang K, Tan T (2008) Robust automated ground plane rectification based on moving vehicles for traffic scene surveillance. In: Proceedings of IEEE international conference on image processing, pp 1364–1367

161. Zhao B, Xing E (2014) Hierarchical feature hashing for fast dimensionality reduction. In: Proceedings of IEEE conference on computer vision and pattern recognition, pp 2051–2058

162. Zheng Y, Shen C, Hartley R, Huang X (2011) Pyramid center-symmetric local binary/trinary patterns for effective pedestrian detection. In: Proceedings of Asian conference on computer vision, pp 281–292

163. Zhou B, Lapedriza A, Xiao J, Torralba A, Oliva A (2014) Learning deep features for scene recognition using Places database. In: Ghahramani Z, Welling M, Cortes C, Lawrence N, Weinberger K (eds) Advances in neural information processing systems 27, Curran Associates, Inc., pp 487–495

164. Zhou H, Hermans T, Karandikar AV, Rehg JM (2010) Movie genre classification via scene categorization. In: Proceedings of the international conference on multimedia, pp 747–750

Chapter 7

Graphical Password Authentication Scheme for Embedded Platform

Siva Janakiraman, K. Thenmozhi, John Bosco Balaguru Rayappan and Rengarajan Amirtharajan

Department of Electronics and Communication Engineering School of Electrical and Electronics SASTRA University, Thanjavur, Tamil Nadu, India

ABSTRACT

Numerous existing graphical systems in the embedded world for password authentication use image, area or pattern selection process. This study presents a novel method that employ image selection at its first level and an isolated pattern selection process on a magnified image display at its next level to provide a well secured password authentication system. The firmware for the proposed authentication scheme was developed under Eclipse open source platform and tested on Cortex-A8 embedded device as a standalone system with the aid of a TFT based graphical LCD screen, a character LCD and a keypad.

KEYWORDS

Learning behavior, Diagnostic feedback, Neural networks, On-line multiple-choice questions

1. INTRODUCTION

Data in digital form can be secured via scrambling process called cryptography or by means of hiding on other cover media known as steganography (Amirtharajan *et al.*, 2012; Amirtharajan and Rayappan, 2013; Amirtharajan *et al.*, 2013a-j; Chan and Cheng, 2004; Cheddad *et al.*, 2010; Praveenkumar *et al.*, 2012a, b, 2013a-d, 2014a-m; Thanikaiselvan *et al.*, 2012a-c, 2013a, b). These security schemes made use of still images and randomness (Amirtharajan and Rayappan, 2012; Janakiraman *et al.*, 2012b; Rajagopalan *et al.*, 2014) as methods for securing the data. Embedded devices such as microcontrollers (Janakiraman *et al.*, 2012a; Janakiraman *et al.*, 2014) and Field Programmable Gate Arrays (FPGAs) (Janakiraman *et al.*, 2013; Rajagopalan *et al.*, 2012a, b) have been the choice of platform for the implementation of security in devices that requires portability. In the increasingly internet-dependent world of today, it is imperative to have sound security and password authentication schemes (Tazawa *et al.*, 2010).

There are two important requirements for an effective password; it must be easy to remember and must be hard to crack. Thus the characteristics of a secure password are:

- Password must be long enough
- Should contain a blend of alphabets (upper case and lower case), numbers as well as special characters
- Should never contain any phrase book words or user's private information

Recent years have seen an exponential increase in the number of transactions being carried out online. Thus the users are required to manage different accounts for the various services that they avail. These days, graphical password is being regarded as promising alternative to replace traditional text-based password in network security (Meng, 2012). Here, the users interact with images for authentication rather than input alpha-numeric strings. This is because image-based password provides a high level of security and is user friendly. The categories of graphical authentication schemes (Hashemi *et al.*, 2012) include image, area or pattern selection. The image schemes involve a click to select an image from the pool; area-based method needs sequence of image selection while pattern-based method relies on drawing some patterns.

People saw the importance of security on the internet and began incorporating techniques to build a stronger secure network. They began to realise the effectiveness of image-based password authentication schemes in providing secure systems. This gradual move towards image-based passwords is seen in the banking sector, e.g., the CUB online banking portal uses image-based password (i.e., where the user selects an image from a group of images displayed) as an entry level authentication and conventional text-based passwords as the second level of authentication. The pattern lock authentication scheme has been incorporated in devices where there is a touch panel interface e.g., smart phones and tablets.

This scheme involves the user drawing a specific continuous pattern i.e., a pattern which is drawn without the user taking his finger off the touch screen. But touch screens are not at everyone's disposal. Thus this study was aimed to design a system that has a wider reach among the customers. Here, we let the user to select an image from a small pool at the first stage. In the next stage it is then combined with pass point method (Wiedenbeck *et al.*, 2005) where, discrete point selection process (click area/click point) is carried out on the selected image. In between the image selection and area selection, the user can select a magnification factor before he starts picking his pass points (areas) on the selected image. The proposed scheme can be effectively put to use among many others, in the areas like banking, finance, online shopping, electronic mail, etc.

2. MATERIALS AND METHODS

2.1 Proposed password authentication scheme

Hardware implementation: The two stage graphical password authentication scheme is implemented on the application level embedded processor CORTEX-A8 where, a graphical TFT based colour graphical LCD (GLCD) screen of dimension 240x320 (RowsxColumns) is interfaced via Serial Peripheral Interface (SPI) to display the required images. A 16x2 character LCD (CLCD) interfaced via Inter Integrated Circuit (I^2C) is used to display the options to be selected by the user for the purpose of password creation, verification or to gain access to the login (authentication). The detailed guide lines for the user in selecting the options displayed in character LCD is provided through serial communication via the on-chip Universal Synchronous Asynchronous Receiver Transmitter (USART) of the embedded processor.

The layout map of GLCD with pixel numbers and row, column positions are given in Fig. 1a. Data is written to the colour GLCD screen by means of writing to the particular memory location using a pointer. The pixel values written to the memory locations contain four bytes of data. The MSB byte, a don't care value is usually kept as zero. The remaining three bytes of data correspond to Red, Green and Blue (RGB) colour components. Grey scale images can be displayed by writing the same eight bit pixel value of the image to all the three fields (RGB) of a particular address as shown in Fig. 1b.

The entire screen is made white (cleared), by writing the value 0xFFFFFFFF to the address corresponding to all pixels of the GLCD screen before the display of actual images. The general principle for the magnified display of 100x100 image as 200x200 image on the GLCD screen is to write the same pixel value in multiple locations in a proportionate manner so that the image appears magnified.

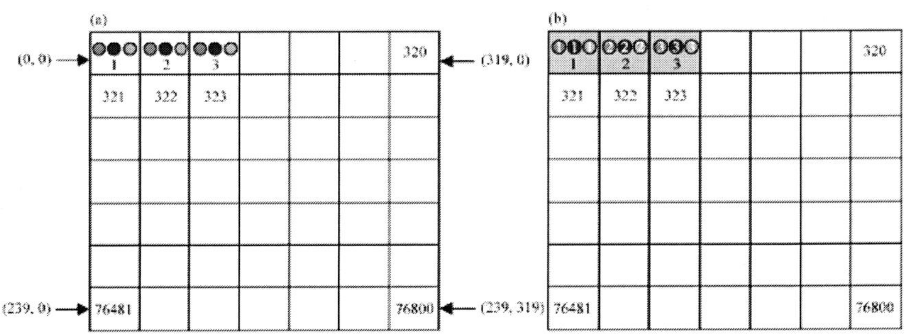

Figure 1 (a-b): (a) Layout of GLCD and (b) Grey image display using RGB components

The square image is made double of its size on both width and height by copying the original pixel values in its next column and row. For example, the first pixel value is written in (1, 1) (1, 2) (2, 1) (2, 2) locations so as to double the image size on both height and width. The grey image display in original and magnified size along with the pixel positions in the display layout to be written with the values of image pixel are shown in Fig. 2.

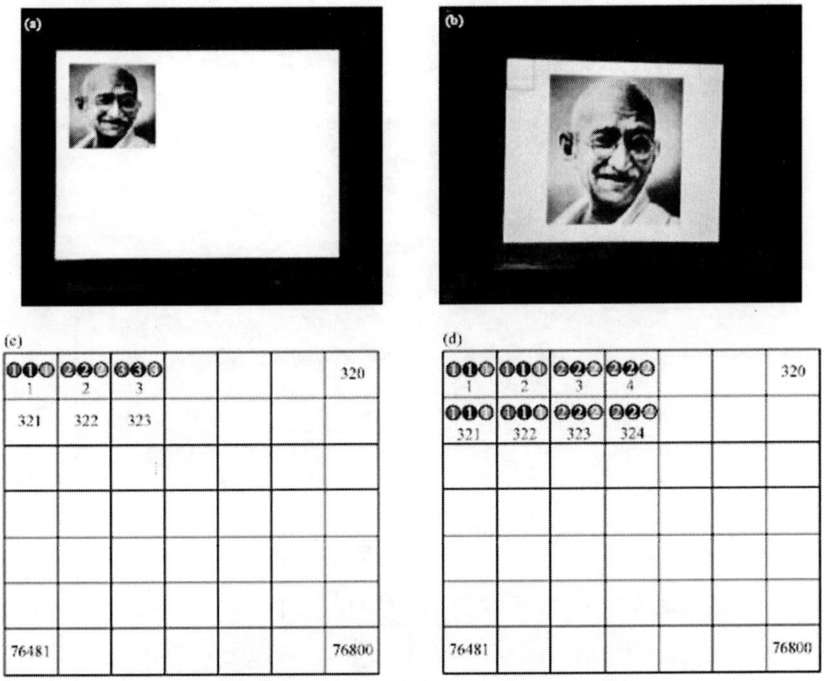

Figure 2(a-d): (a) Original grey image 100x100, (b) Magnified grey image 200x200, (c) Pixel layout for original image display and (d) Pixel layout for magnified image display

The flowchart of the overall authentication scheme is shown in Fig. 3. The various option selections can be done using the 6x6 keypad connected via a Power Management Integrated Chip (PMIC) with built-in scanning mechanism for the detection of hardware key press.

Password creation: Initially the users are asked to identify him as a new user or an existing user. A new user is asked to set a password while an existing user is directed to the verification stage. Every new user is first assigned with an account number that acts as user name. For the first stage of password setting, the new user has to select an image as password from the four 100x100 images displayed on the graphical LCD screen. Then, the selected image is magnified and displayed as a 200x200 image on the screen. There is also a 40x40 selection box appears on the top left corner of the screen for the user to select specific areas of the image.

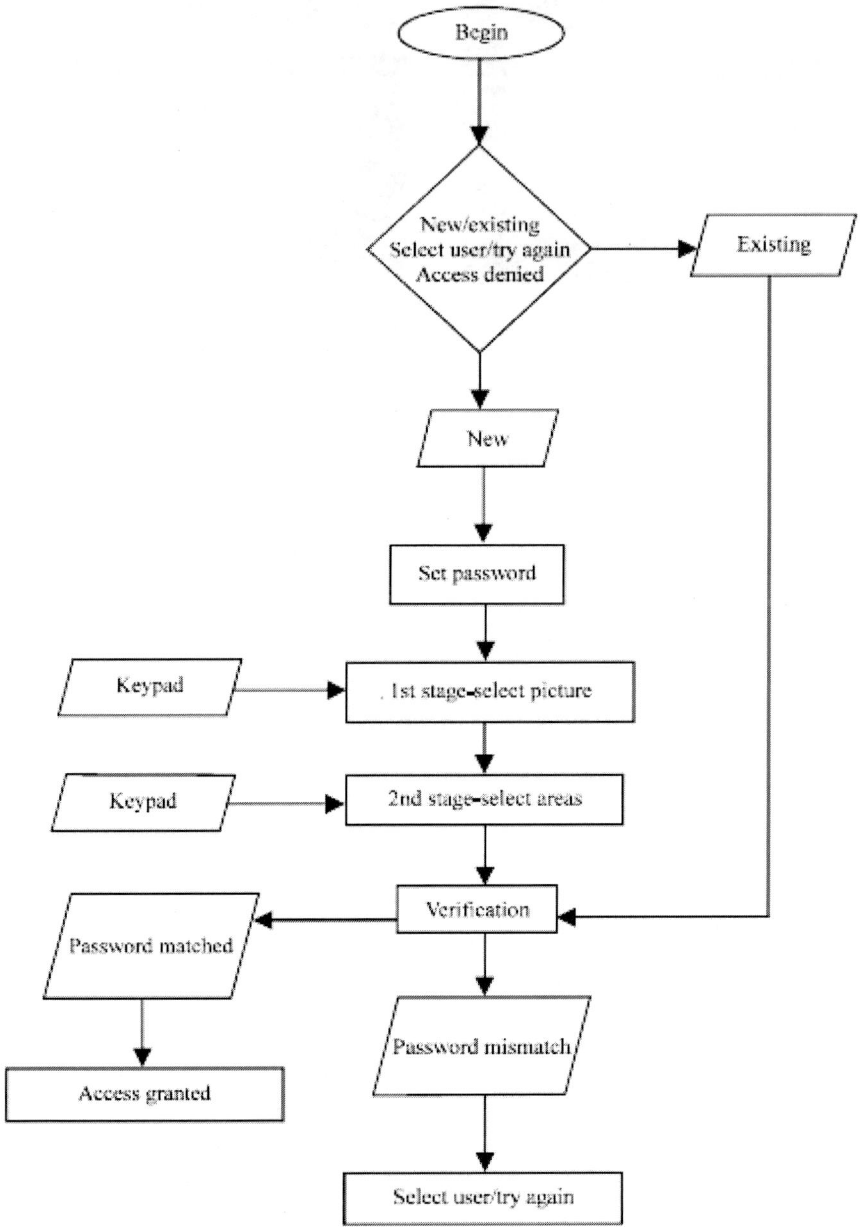

Figure. 3: Flowchart for overall authentication scheme

The second stage of password setting involves the user moving the selection box across the magnified image and selecting portions of the image in any specific order as he desires.

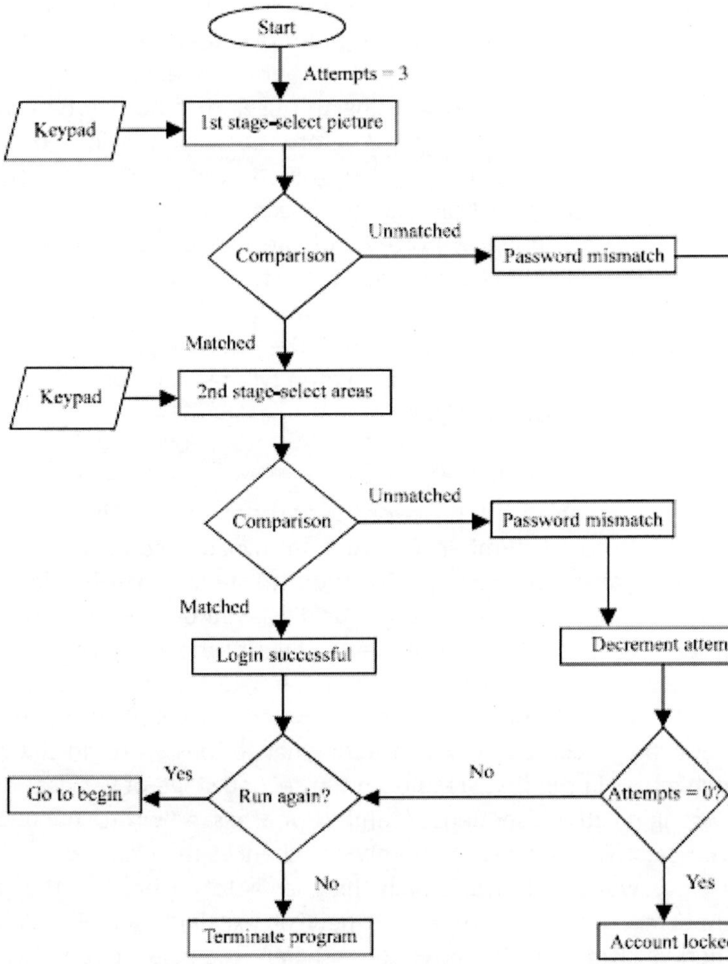

Figure 4: Flowchart for verification process

The maximum number of areas that can be selected on the image by the user (password length) is four. The selection box is moved with the help of keypad inputs 4 for left, 8 for up, 6 for right, 2 for down and 5 for selecting that portion of the image as one digit of the password. In case of mistake in the selection of the desired area, the user is given an option to deselect the last selected area. After the user completes setting the two levels of passwords, a success message is displayed on the character LCD.

Authentication/verification: The first stage of authentication involves the user selecting the correct image from a group of images displayed on the GLCD screen. When the user selects a wrong image, a "Password Mismatch" message is displayed warning the user that he has two more chances to type in the correct password before the account is locked. Once the user selects the correct image, a magnified version of the image is displayed along with the selection box on the screen. The user then has to select, in the correct sequence, portions of the image that he set as his password using the selection box. In case of mistake in the selection of the desired area, the user is given an option to deselect the last selected area. After selecting all the digits of the password, the user can give the required keyboard input to state that he has completed typing in the password. The verification process is shown in Fig. 4.

Initially the user identifies him as an existing user or new user. Existing users are directly taken to the password verification phase. The user then has to input his account number. Entering an invalid account number by the user displays an error message. For the first stage of verification, the user has to select one image from the four displayed on the GLCD screen. The user then proceeds to the second level of verification, where he has to select the areas in correct sequence from the selected image. Once the user enters the full password, a comparison process takes place where the entered password is compared with the original set password. In the case of mismatch either in the first stage or in the second stage of verification in terms of wrong positions/sequence/number of areas, a "wrong password" message appears. This detects the number of chances the user has to enter the correct password and after which the user is taken back to the start screen where he can again enter as a new or existing user. Finally, the account gets locked for the next 24 h when the user enters wrong password for three consecutive times. A "Password successful" message appears in the case of a perfect match and access to his account is granted.

Restrictions: The program does not allow the users to select the background white spaces as password characters and thus displays "Invalid selection" message for the same in CLCD. It also prompts the user to select the password area again. This program allows the user to select a minimum of one area and a maximum of four areas as password. Thus, after every area selected, it prompts the user as to whether he wants to select the next area as his password. Whenever a user tries to set a password that is more than four characters in length, an error message is displayed and the program considers the first four areas selected by the user as the password. Once the two level of password setting is completed, a success message appears on the CLCD after which the user enters the verification phase. The user is also given the option to change password after every successful login.

3. RESULTS AND DISCUSSION

The maximum password spaces that can be supported by the scheme can be calculated as follows:

Size of the image = m×n and Size of grid square = r×s

No. of non-overlapping grid squares available in the image, 'g'= (m×n)/(r×s)

No. of passwords spaces when all the users are allowed to select the password length 'k' (k-clicks) in a image with 'g' grids is given as g^k (1)

Total passwords spaces obtainable when the users are allowed to select the minimum password length k = 1 and maximum length of k in a image with 'g' grids where, k≥1 can be given as: $g^1+g^2+\ldots+g^k$ (2)

The sample results for pool of 4 images with sizes 100x100 each and selected image displayed in size 200x200 by the magnification factor of 2 with grid square (selection box) of size 40x40 in its top right corner displayed in the GLCD screen during the first stage, area selection in the second stage and final password confirmation message obtained in CLCD at the second stage are shown in Fig. 5.

Features and factors of improvement: The popular pattern lock feature can only be used in smart phones and tablets i.e., where a touch screen is being used and so does not extend to basic phones. The image based password authentication scheme proposed here brings a solution to the problem and helps strengthen the security even in a low cost device with monochrome/colour display and keypad. Here, this method used an application level embedded processor that can access huge memory to support the image and password storage to manage multiple users. The use small sized images of grey format, reduces the memory requirement. The magnified display of selected images during the area selection process increases the potential number of areas that can be selected which in turn elevates the security and the possible number of users. Though this method fix a restriction on the maximum length of the password, the flexibility given by the algorithm for the users to select variable length for their password helps in accommodating as many users as possible. Reduction in the original size of the images stored can help minimizing memory requirement. More users can be lodged with few images by means of increasing the image magnification factor. The size of the selection box can be made still smaller which will increase the number of areas that can be selected. This in turn increases the number of password combinations and reduces threat of attack to a great extent. When the number of selection areas that a user can choose from increases, the system becomes capable of supporting a larger pool of users with non-overlapping passwords with a restriction on maximum number of areas to be selected as password.

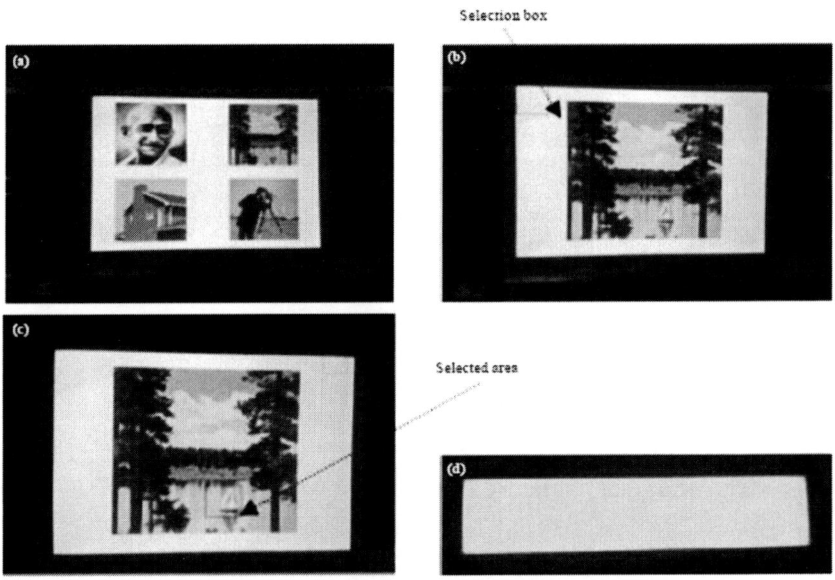

Figure 5(a-d): (a) Image pool in first stage, (b) Selected image in magnified scale with selection box, (c) Area selection in magnified image and (d) Message after password matching

Table 1 compares the use of various features and mechanisms between the proposed method and previous method (Wiedenbeck *et al.*, 2005).

Table 1. Comparison result on feature/mechanism of the proposed method

Features/mechanism	Pass points on embedded devices (proposed)	Pass points on PC (Wiedenbeck *et al.*, 2005)
Target platform	ARM Cortex-A8	PENTIUM- IV PC
Screen size used	3.5"	19"
Input device	Keypad	Mouse
Target application	Password authentication in standalone	
Embedded devices	Password authentication in PC	
Factors deciding the password space	Image size Grid square size Password length (max) Magnification factor	Image size Grid square size Password length (max)
Click points	Square well defined boundaries	Grid square boundaries with tolerance
Preliminary verification before actual password entry	Yes (image selection from the pool of images displayed on the screen)	No (single user image will be displayed)
Password length	Variable (within the maximum specified by the scheme)	Fixed (all users must have equal length)
Grid square size	Variable	Variable
Magnification factor	Variable (selectable by user from available options)	Fixed (pre-determined in the scheme)

Table 2. Comparison on password space and memory requirement

Method	Image size	Image type	Memory required for image storage	Grid square size	Magnification factor	No. of grid squares	Maximum/minimum length of password	Size of password space (given by Eq. 1 and 2)
Proposed	100×100	Gray	9.76 KB	20×20	2	100	4/1	101.1001×10^5
Wiedenbeck *et al.* (2005)	100×100	RGB	29.29 KB	20×20	1	50	4/4	6.25×10^5

The Table 2 shows a significant improvement in size of the password space and reduction in the space required for image storage by the proposed method in comparison with earlier method by Wiedenbeck *et al.* (2005) when using equal sizes of image, grid square and maximum password length (clicks).

4. CONCLUSION

In the proposed image and area selection scheme the pixel values of the selected areas in the magnified image are compared for password authentication. This makes the image magnification factor used in the scheme to act as a novel security key. Thus, this scheme combines both authentication at image level and authentication at area level inside an image (pixel values) to provide a high level of security on embedded platform. Using the keypad and GLCD for the password selection makes the scheme also suitable for low end embedded devices for single user authentication without much demand on memory.

ACKNOWLEDGMENT

The authors wish to acknowledge SASTRA University for providing infrastructural support to carry out this study.

REFERENCES

1. Amirtharajan, R. and J.B.B. Rayappan, 2012. Pixel authorized by pixel to trace with SFC on image to sabotage data mugger: A comparative study on PI stego. Res. J. Inform. Technol., 4: 124-139.

2. Amirtharajan, R. and J.B.B. Rayappan, 2013. Steganography-time to time: A review. Res. J. Inform. Technol., 5: 53-66.

3. Amirtharajan, R., G. Devipriya, V. Thanikaiselvan and J.B.B. Rayappan, 2013. High capacity triple plane embedding: A colour stego. Res. J. Inform. Technol., 5: 373-382.

4. Amirtharajan, R., J. Qin and J.B.B. Rayappan, 2012. Random image steganography and steganalysis: Present status and future directions. Inform. Technol. J., 11: 566-576

5. Amirtharajan, R., K. Karthikeyan, M. Malleswaran and J.B.B. Rayappan, 2013. Kubera kolam: A way for random image steganography. Res. J. Inform. Technol., 5: 304-316.

6. Amirtharajan, R., K.M. Ashfaaq, A.K. Infant and J.B.B. Rayappan, 2013. High performance pixel indicator for colour image steganography. Res. J. Inform. Technol., 5: 277-290.

7. Amirtharajan, R., M.V. Abhiram, G. Revathi, J.B. Reddy, V. Thanikaiselvan and J.B.B. Rayappan, 2013. Rubik's cube: A way for random image steganography. Res. J. Inform. Technol., 5: 329-340.

8. Amirtharajan, R., P. Archana and J.B.B. Rayappan, 2013. Why image encryption for better steganography. Res. J. Inform. Technol., 5: 341-351.

9. Amirtharajan, R., P.S. Priya and J.B.B. Rayappan, 2013. Pixel indicated user indicator: A muxed stego. Res. J. Inform. Technol., 5: 73-86.

10. Amirtharajan, R., R. Subrahmanyam, J.N. Teja, K.M. Reddy and J.B.B. Rayappan, 2013. Pixel indicated triple layer: A way for random image steganography. Res. J. Inform. Technol., 5: 87-99.

11. Amirtharajan, R., S. Sulthana and J.B.B. Rayappan, 2013. Seeing and believing is a threat: A visual cryptography schemes. Res. J. Inform. Technol., 5: 435-441.

12. Amirtharajan, R., S.D. Roy, N. Nesakumar, M. Chandrasekar, R. Sridevi and J.B.B. Rayappan, 2013. Mind game for cover steganography: A refuge. Res. J. Inform. Technol., 5: 137-148.

13. Amirtharajan, R., V. Rajesh, P. Archana and J.B.B. Rayappan, 2013. Pixel indicates, standard deviates: A way for random image steganography. Res. J. Inform. Technol., 5: 383-392.

14. Chan, C.K. and L.M. Cheng, 2004. Hiding data in images by simple LSB substitution. Pattern Recognit., 37: 469-474.

15. Cheddad, A., J. Condell, K. Curran and P. Mc Kevitt, 2010. Digital image steganography: Survey and analysis of current methods. Signal Process., 90: 727-752.

16. Hashemi, M., N. Ithnin and R. Pakdel, 2012. Multi touch graphical password: Usability features. Asian J. Applied Sci., 5 : 20-32.

17. Janakiraman, S., K.V.S.K. Kumar, R.R.K. Reddy, A. Srinivasulu, R. Amirtharajan, K. Thenmozhi and J.B.B. Rayappan, 2014. Humming bird with coloured wings: A feedback security approach. Inform. Technol. J., 13: 2022-2026.

18. Janakiraman, S., R. Amirtharajan, K. Thenmozhi and J.B.B. Rayappan, 2012. Pixel forefinger for gray in color: A layer by layer stego. Inform. Technol. J., 11: 9-19.

19. Janakiraman, S., R. Amirtharajan, K. Thenmozhi and J.B.B. Rayappan, 2012. Firmware for data security: A review. Res. J. Inform. Technol., 4: 61-72.

20. Janakiraman, S., S. Rajagopalan, K. Thenmozhi, H.N. Upadhyay and J. Ramanathan *et al.*, 2013. Captivating CODEC Stego (CCS): A cover on camouflage. Res. J. Inform. Technol., 5: 160-170.

21. Meng, Y., 2012. Designing click-draw based graphical password scheme for better authentication. Proceedings of the 7th International Conference on Networking, Architecture and Storage, June 28-30, 2012, Xiamen, Fujian, pp: 39-48.

22. Praveenkumar, P., G. Ashwin, S.P.K. Agarwal, S.N. Bharathi, V.S. Venkatachalam, K. Thenmozhi and R. Amirtharajan, 2014. Rubik's cube blend with logistic map on RGB: A way for image encryption. Res. J. Inform. Technol., 6: 207-215.

23. Praveenkumar, P., G.S. Hemalatha, B. Reddy, K. Thenmozhi, J.B.B. Rayappan and R. Amirtharajan, 2014. Secret link through simulink: A stego on OFDM channel. Inform. Technol. J., 13: 1999-2004.

24. Praveenkumar, P., K. Thenmozhi, J.B.B. Rayappan and R. Amirtharajan, 2014. Purposeful error on OFDM: A secret channel. Inform. Technol. J., 13: 1985-1991.

25. Praveenkumar, P., K. Thenmozhi, J.B.B. Rayappan and R. Amirtharajan, 2014. Stego in multicarrier: A phase hidden communication. Inform. Technol. J., 13: 2011-2016.

26. Praveenkumar, P., K. Thenmozhi, J.B.B. Rayappan and R. Amirtharajan, 2014. Inserted embedding in OFDM channel: A multicarrier stego. Inform. Technol. J., 13: 2017-2021.

27. Praveenkumar, P., K. Thenmozhi, J.B.B. Rayappan and R. Amirtharajan, 2014. Data puncturing in OFDM channel: A multicarrier stego. Inform. Technol. J., 13: 2037-2041.

28. Praveenkumar, P., K. Thenmozhi, J.B.B. Rayappan and R. Amirtharajan, 2014. Spread and hide-a stego transceiver. Inform. Technol. J., 13: 2061-2064.

29. Praveenkumar, P., K. Thenmozhi, M.N. Dinesh and R. Amirtharajan, 2013. Fixing, padding and embedding: A modulated stego. Int. J. Eng. Technol., 5: 2257-2261.

30. Praveenkumar, P., M. Nagadinesh, P. Lakshmi, K. Thenmozhi, J.B.B. Rayappan and R. Amirtharajan, 2013. Convolution and viterbi EN (DE) coders on OFDM hides, rides and conveys message-A neural STEGO. Proceedings of the International Conference on Computer Communication and Informatics, January 4-6, 2013, Coimbatore, pp: 1-5.

31. Praveenkumar, P., R. Amirtharajan, K. Thenmozhi and J.B.B. Rayappan, 2012. Phase for face saving-a multicarrier stego. Proc. Eng., 30: 790-797.

32. Praveenkumar, P., R. Amirtharajan, K. Thenmozhi and J.B.B. Rayappan, 2012. Regulated OFDM-role of ECC and ANN: A review. J. Applied Sci., 12: 301-314.

33. Praveenkumar, P., R. Amirtharajan, K. Thenmozhi and J.B.B. Rayappan, 2013. Can we reduce PAPR? OFDM+PTS+SLM+STEGO: A novel approach. Asian J. Sci. Res., 6: 38-52.

34. Praveenkumar, P., R. Amirtharajan, K. Thenmozhi and J.B.B. Rayappan, 2013. OFDM with low PAPR: A novel role of partial transmit sequence. Res. J. Inform. Technol., 5: 35-44.

35. Praveenkumar, P., R. Amirtharajan, K. Thenmozhi and J.B.B. Rayappan, 2014. Sub carriers carry secret: An absolute stego approach. J. Applied Sci., 14: 1728-1735.

36. Praveenkumar, P., R. Amirtharajan, K. Thenmozhi and J.B.B. Rayappan, 2014. Double layer encoded encrypted data on multicarrier channel. J. Applied Sci., 14: 1689-1700.

37. Praveenkumar, P., R. Amirtharajan, K. Thenmozhi and J.B.B. Rayappan, 2014. Coded crypted converted hiding (C^3H)-a stego channel. J. Applied Sci., 14: 1786-1797.

38. Praveenkumar, P., R. Amirtharajan, R.S. Janani, K. Thenmozhi and J.B.B. Rayappan, 2014. Multi (Carrier+Modulator) adaptive system-an anti fading stego approach. J. Applied Sci., 14: 1836-1843.

39. Praveenkumar, P., R. Deepak, K. Thenmozhi, J.B.B. Rayappan and R. Amirtharajan, 2014. Reversible steganography on OFDM channel: A role of cyclic codes. Inform. Technol. J., 13: 2047-2051.

40. Praveenkumar, P., R. Hemalatha, R. Uma, K. Madhunisha, K. Thenmozhi and R. Amirtharajan, 2014. Image $^{Zoning}\rightarrow$ encryption. Res. J. Inform. Technol., 6: 368-378.

41. Rajagopalan, S., R. Amirtharajan, H.N. Upadhyay and J.B.B. Rayappan, 2012. Survey and analysis of hardware cryptographic and steganographic systems on FPGA. J. Applied Sci., 12: 201-210.

42. Rajagopalan, S., S. Janakiraman, H.N. Upadhyay and K. Thenmozhi, 2012. Hide and seek in silicon: Performance analysis of Quad block Equisum Hardware Steganographic systems. Procedia Eng., 30: 806-813.

43. Rajagopalan, S., Y. Ravishankar, H.N. Upadhyay, J.B.B. Rayappan and R. Amirtharajan, 2014. Modeling combo PR generator for Stego Storage Self Test (SSST). Inform. Technol. J., 13: 1936-1944.

44. Tazawa, H., T. Katoh, B.B. Bista and T. Takata, 2010. A user authenticaion scheme using multiple passphrases and its arrangement. Proceedings of the International Symposium on Information Theory and its Applications, October 17-20, 2010, Taichung, pp: 554-559.

45. Thanikaiselvan, V., K. Santosh, D. Manikanta and R. Amirtharajan, 2013. A new steganography algorithm against chi square attack. Res. J. Inform. Technol., 5: 363-372.

46. Thanikaiselvan, V., P. Arulmozhivarman, J.B.B. Rayappan and R. Amirtharajan, 2012. Graceful graph for graceful security-towards a STE (G) Raph. Res. J. Inform. Technol., 4: 220-227.

47. Thanikaiselvan, V., P. Arulmozhivarman, R. Amirtharajan and J.B.B. Rayappan, 2012. Horse riding and hiding in image for data guarding. Proc. Eng., 30: 36-44.

48. Thanikaiselvan, V., P. Arulmozhivarman, R. Amirtharajan and J.B.B. Rayappan, 2012. Wavelet Pave the Trio travel for a secret mission: A stego vision. Global Trends Inform. Syst. Software Applic., 270: 212-221.

49. Thanikaiselvan, V., P. Arulmozhivarman, S. Subashanthini and R. Amirtharajan, 2013. A graph theory practice on transformed image: A random image steganography. Sci. World J., Vol. 2013 10.1155/2013/464107

50. Wiedenbeck, S., J. Waters, J.C. Birget, A. Brodskiy and N. Memon, 2005. PassPoints: Design and longitudinal evaluation of a graphical password system. Int. J. Hum. Comput. Stud., 63: 102-127.

Chapter 8

Intelligent diagnostic feedback for online multiple-choice questions

R. Guo, D. Palmer-Brown, S. W. Lee, F. F. Cai

Faculty of Life Sciences and Computing, London Metropolitan University, Holloway Road, London N7 8DB, UK

ABSTRACT

When students attempt multiple-choice questions (MCQs) they generate invaluable information which can form the basis for understanding their learning behaviours. In this research, the information is collected and automatically analysed to provide customized, diagnostic feedback to support students' learning. This is achieved within a web-based system, incorporating the snap-drift neural network based analysis of students' responses to MCQs. This paper presents the results of a large trial of the method and the system which demonstrates the effectiveness of the feedback in guiding students towards a better understanding of particular concepts.

KEYWORDS

Learning behavior, Diagnostic feedback, Neural networks, On-line multiple-choice questions

1. INTRODUCTION

In recent years, e-learning has become commonplace in higher education. The involvement of intelligent e-learning systems has the potential to make higher education accessible with increasing convenience, efficiency and quality of study. According to the Hefce National Student Survey (2007–2010), in England, only about half of students believe that: (1) feedback on their work has been prompt; (2) feedback on their work has helped to clarify things they did not understand; (3) they have received detailed comments on their work. These reports reveal that the feedback and its related fields are one of the weakest areas in higher education in England. This research investigates the relative effectiveness of different types of feedback, and how to optimize feedback to facilitate deep learning. It compares and contrasts several methods in order to investigate the effectiveness of using intelligent feedback towards modeling the stages of students' knowledge. The investigation will lead to an understanding of the potential of the on-line diagnostic feedback across different subject areas.

The virtual learning environment (VLE) as a an e-learning system based on web which can model personal education by providing virtual access to classes, concepts, key learning point, tests, homework, grades, assessments, and other external resources. It can also provide a method to help students to enhance their learning experiences, and to help teachers manage the gap between teaching and learning. The VLE presented in this paper provides a generic method for intelligent analysis and grouping of student responses that applicable to any area of study. This tool offers important benefits: immediate feedback, significant time-saving evaluating assignments, and consistency in the learning process. The time taken to create the feedback is well spent not only because this feedback can be reused, but also it is made available through the system to large numbers of students.

2. BACKGROUND AND REVIEW OF PREVIOUS WORK

Ma (2006) described intelligent tutoring systems (ITSs) as the milestone of the advanced generation of computer-aided instruction systems, and concluded their key feature as 'the ability to provide a user-adapted presentation of the teaching material'. Rane and Sasikumar (2007) pointed out that to overcome the lack of the presence of a teacher, intelligent tutoring systems attempt to simulate a teacher, who can guide the student's study based on the student's level of knowledge by giving intelligent instructional feedback. Furthermore, according to Blessing et al. (2007), the intense interaction and feedback achieved by intelligenttutoring systems can significantly improve student learning gains. In addition, in Gheorghiu and Vanlehn (2008)'s paper, they also suggested that meaningful, constructive and adaptive feedback is the essential feature of ITSs, and it is such feedback that helps students achieve strong learning gains.

Cullen et al. (2002, cited by Heinze et al. 2007) suggest that feedback is one of the most important indicators of good education. Little (2001) also presents that providing feedback, which can continuously reinforce learning progress and promote learners' attention and engagement, is crucial to effective learning. Besides, Vasilyeva et al. (2008) also state that 'the design of feedback is a critical issue of online assessment development within web-based learning systems'. However, reviewing the education reality, it is surprising that as one of the most important communication methods between teachers and students, feedback was one of the weakest areas according to the Hefce National Student Survey (2007–2010) in the UK. Thus, how to design an optimized intelligent feedback system becomes a major and critical problem to develop a successful e-learning system.

To make the feedback effective and meaningful, a range of quality attributes need to be achieved. Hatziapostolou and Paraskakis (2010) summarized the work by Race (2006), Irons (2008) and Juwah et al. (2004) and suggested that in order to improve learning gains, formative feedback should address as many as possible of the following attributes, including constructive, motivational, personal, manageable, timely and directly related to assessment criteria and learning outcomes. (1) Constructive. Effective feedback should be constructive, as constructive feedback can lead to more thinking and cognitive learning which in turn improve the student's learning (Bang 2003 and Alessi and Trollip 2001). As Nelson and Schunn (2009) argued, effective feedback should be able to guide a learner 'to change performance in a particular direction rather than just towards or away from a prior behavior'. (2) Motivational. Effective feedback should be motivational to empower and encourage students to learn more, as feedback can affect students' feelings and attitudes towards study, which in turn affect their engagement in the learning process, (Juwah et al. 2004, cited by Hatziapostolou and Paraskakis 2010). (3) Garber (2004) also argued, 'the more personalized the feedback becomes, the more meaning it can have for the individual receiving it……and the more likely the individual will be receptive to the feedback……If a person does not believe in the reliability or validity of the feedback, it will have little or no benefit'. (4) Manageable. Effective feedback should be detailed enough to ensure that students clearly understand their strengths and weaknesses, and have enough materials to guide them to achieve the learning goals. At the same time, the feedback should not be over-detailed to avoid confusing, and make students can easily interpret it and get the point (Hatziapostolou and Paraskakis2010). (5) Timely. Feedback should be delivered timely, as students can more easily utilize feedback when they can still remember how they just processed the task (Race 2006), and the reasoning that led to the error is still accessible (Reiser and Kimberg 1992). As Anderson, Boyle and Reiser (1985, cited by Reiser and Kimberg 1992) argued, tutors should provide immediate feedback to students, as 'the learning mechanism for adjusting a faulty rule or forming a new correct rule relies upon the problem situation being active in memory'. (6) Directly related to assessment criteria/learning outcomes. Effective feedback should be directly related to assessment criteria/learning outcomes so that it can explain students' achievement towards the intended learning outcomes, knowledge gaps and specific errors (Hatziapostolou and Paraskakis2010). Thus, the students can be guided and adjust their effort to achieve the intended leaning outcomes (Race and Brown 2005). Additionally, Clark and Mayer (2009) further argued that learning goal oriented feedback is more effective than performance goal oriented feedback. In another word, feedback should be designed to inform the learners their progress toward achieving a learning goal rather than compare a learner performance with other learners'.

In addition to the required attributes above, many researchers also mentioned that various methods should be used in feedback to ensure better perception of feedback. For example, Özdener and Satar (2009) suggested using animation techniques to achieve better reception and perception of the feedback.

Springgay and Clarke (2007) suggested including examples in feedback to achieve better perception of feedback. Multiple choice questions (MCQs) is an effective way to provide students with feedback. The use of multiple-choice questions has been widely studied. A number of advantages can be found in the Epstein et al. (2002), Higgins and Tatham (2003) and Kuechler and Simkin (2003) researches: rapid feedback, automatic evaluation, perceived objectivity, easily computed statistical analysis of test results and the reuse of questions from databases as required, thus saving time for instructors. On the other hand there are also some researches (e.g. Paxton 2000) shows that MCQs have some disadvantages: significant effort is required to construct MCQs, they only evaluate knowledge and recall, and they are unable to test literacy and creativity. Although the MCQs have been primarily used for summative evaluation, they also serve formative assessment purposes. Formative assessment provides students with feedback that highlights areas for further study and indicates the degree of progress. There are many studies investigating the role of different type of feedback in Web-based assessments that report positive results from the use of MCQs in online tests for formative assessments.

Many researches investigating the effect of different types of feedback in web-based assessments showed the positive results of using MCQs in online test for formative assessment (e.g. Epstein et al. 2002; Higgins and Tatham 2003; Kuechler and Simkin 2003; Payne et al. 2007). Higgins and Tatham (2003) studied the use of MCQs in formative assessment in a web-based environment using WebCT for a level one unit on undergraduate law degree. They summed that they could forecast all the possible errors for a question and write a general feedback for this question. However, using this type of feedback, it could be difficult to predict all the possible errors and produce the general feedback for a combination of questions, and it would be impossible for a large test banks (e.g. 3 questions with 5 answers would require 125 answer combinations; 5 questions with 5 answers require 3,125 combinations, etc.).

Payne et al. (2007) assessed the effectiveness of three different forms of feedback (corrective, corrective explanatory, and video feedback) used in e-learning to support students' learning. This type of feedback shows exactly which questions are answered correctly or not, with further corrective explanation and video feedback. Our approach to feedback is different from the above. The intelligent diagnostic feedback we present is concept-oriented instead of question-oriented. The learners are encouraged to review the concepts they misunderstood through the feedback in order to retake the test again and study further. It is important that each category of answers is associated with carefully designed feedback based on the level of understanding and prevalent misconceptions of that category-group of students so that every individual student can reflect on his or her learning level using this diagnostic feedback. In addition, when students retake the test they receive new feedback according to his or her knowledge state, which in turn leads to more self-learning. Moreover, concept-based feedback can also prevent the student from guessing the right answers; if the students do not read the diagnostic feedback carefully, they may not even know which questions were answered incorrectly. In conclusion the focus of our current research is to combine MCQs and formative online assessment using an intelligent agent to analyse the students' response in order to provide diagnostic feedback. To this end, we deploy a neural network

3. MULTIPLE-CHOICE QUESTIONS ONLINE FEEDBACK SYSTEMS (M-OFS)

To analyse the students' answers, and integrate over a number of questions to gain insights into the students' learning needs, a snap-drift neural network (SDNN) approach is proposed. SDNN provides an efficient means of discovering a relatively small and therefore manageable number of groups of similar answers. In the following sections, an e-learning system based on SDNN is described.

3.1 Snap-drift neural networks (SDNNs)

The learning process involves a combination of fast, convergent, minimalist learning (snap) and overall feature averaging (drift) to capture both precise sub-features in the data and more general holistic features. Snap and drift learning phases are combined within a learning system that toggles its learning style between the two modes. On presentation of input data patterns at the input layer, the distributed SDNN (dSDNN) learns to group them according to their features using snap-drift (Palmer-Brown and Jayne 2011). The neurons whose weight vectors result in them receiving the highest activations are adapted. Weights are normalised weights so that in effect only the angle of the weight vector is adapted, meaning that a recognised feature is based on a particular ratio of values, rather than absolute values. The output winning neurons from dSDNN act as input data to the selection SDNN (sSDNN) module within performs feature grouping and this layer is also subject to snap-drift learning.

The learning process is unlike error minimisation and maximum likelihood for SDNN toggles its learning mode to find a set of sub-features and average feature in the data and uses them to group the data into categories. Each weight vector is bounded by snap and drift: snapping gives the angle of the minimum values (on all dimensions) and drifting gives the average angle of the patterns grouped under the neuron. Snap creates a feature common to all the patterns in the group and gives a high probability of rapid (in terms of epochs) convergence (both snap and drift are convergent, but snap is faster). Drifting, which uses learning vector quantization (LVQ), tilts the vector towards the centroid angle of the group and ensures that an average, generalised feature is included in the final vector. The angular range of the pattern-group membership depends on the proximity of neighbouring groups (natural competition), and can be controlled by adjusting a threshold on the weighted sum of inputs to the neurons.

3.2 Training neural network

The e-learning snap-drift neural network (ESDNN) is trained with the students' responses to questions on a particular topic in a course. The responses are obtained from the previous cohorts of students. Before training, each of the responses from the students is encoded into binary form in preparation for presentation as input patterns for ESDNN. Table 1 shows examples of a possible format of questions for five possible answers and some encoded responses. This version of ESDNN is a simplified unsupervised version of the snap-drift algorithm (Palmer-Brown and Jayne 2011) as shown in Fig. 1.

Table 1. Example of input patterns for ESDNN

Codification	A:00001	B:00010	C:00100	D:01000	N/A:00000
Response	Recorded response				
[C, D, B, A]	[0, 0, 1, 0, 0, 0, 1, 0, 0, 0, 0, 0, 0, 1, 0, 0, 0, 0, 0, 1]				
[E, B]	[1, 0, 0, 0, 0, 0, 0, 0, 1, 0]				
[D, A, A]	[0, 1, 0, 0, 0, 0, 0, 0, 0, 1, 0, 0, 0, 0, 1]				
[A, C, D, B, A]	[0, 0, 0, 0, 1, 0, 0, 1, 0, 0, 0, 10, 0, 0, 0, 0, 0, 1, 0, 0, 0, 0, 0, 1]				

During training, on presentation of an input pattern at the input layer, the dSDNN will learn to group the input patterns according to their general features. In this case, 5 F12 nodes, whose weight prototypes best match the current input pattern, with the highest net input are used as the input data to the sSDNN module for feature classification. In the sSDNN module, a quality assurance threshold is introduced. If the net input of an sSDNN node is above the threshold, the output node is accepted as the winner; otherwise a new uncommitted output node will be selected as the new winner and initialised with the current input pattern. For example, for one group, every response might have in common the answer C to question 2, the answer D to question 3, the answer A to question 5, the answer A to question 6, the answer B to question 8, and the answer A to question 10. The other answers to the other questions will vary within the group, but the group is formed by the neural network based on the commonality between the answers to some of the questions (four of them in that case). From one group to another, the precise number of common responses varies in theory between 1 and X, where X is the number of questions. In this experiment, where there are 10 questions in 1st English trial, the groups had between 5 and 8 (Trial 1) common answers. More details of the steps that occur in ESDNN and the ESDNN learning algorithm are given in (Palmer-Brown and Jayne 2011). The training relies upon having representative training data. The number of responses required to train the system so that it can generate the states of knowledge varies from one domain to another. When new responses create new groups, more training data is required. Once new responses stop creating new groups, it is because those new responses are similar to previous responses, and sufficient responses to train the system reliably are already available. The number of groups formed depends on the variation in student responses.

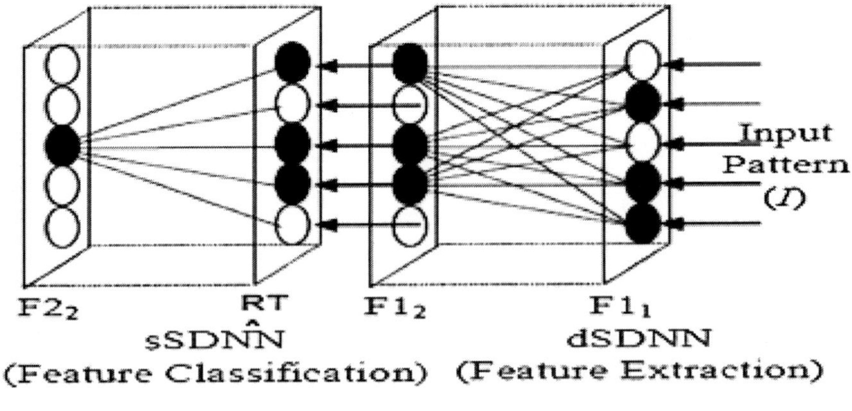

Figure 1. E-learning SDNN architecture

3.3 How the system guides learning

The feedback is designed by academics so that it does not identify which questions were incorrectly answered. The academics are presented with the groups in the form of templates of student responses. Table 2 shows some examples of the group formed. For example, "A/D B mix" represents group 1 characterized by all the students answering A or D to question 1, B to question 2, and mixed answers to question 3. Hence, the educator can easily see the common mistakes in the groups of the student answers highlighted by the tool. The feedback texts are associated with each of the pattern groupings and are composed to address misconceptions that may have caused the incorrect answers common to that pattern group. The student responses, recorded in the database, can be used for monitoring the progress of the students and for identifying misunderstood concepts that can be addressed in subsequent face-to-face sessions. The collected data can be also used to analyze how the feedback influences the learning of individual students by following a particular student's progress over time and observing how that student's answers change after reading the feedback. Student responses can also be used to retrain the neural network and see whether refined groupings are created, which can be used by the educator to improve the feedback. Once designed, MCQs and feedbacks can be reused for subsequent cohorts of students.

Table 2. Example of answer groups

Group number	Group formed		
	Question 1	Question 2	Question 3
1	A/D	B	Mix
2	C	D	D
3	Mix	A	A/B/D

4. EXPERIMENTAL ENVIRONMENT

In order to evaluate the performance and effectiveness of this novel e-learning system, abundant target-oriented testing needs to be carried out in different fields. Furthermore, we also aim to enhance this system to overcome its deficiencies during practical applications. Thus, this study is composed of three main parts. Firstly, we evaluated the system by collecting and analysing a large number of testing data reflecting the students' learning gains by using this system as well as the survey and interview data reflecting the students' satisfaction and attitudes towards this system. Secondly, the investigation will lead to an understanding of the potential of the on-line diagnostic feedback approach across different subject areas. Furthermore, this research should also produce guidelines for the design principles of on-line MCQs in the context of diagnostic feedback learning environments. Thirdly, we managed to enhance the existing system according to the evaluation data. The details of this experiment which are conducted to assess the use of M-OFS during academic year 2010–2011 are shown in next section.

4.1 Data collection and feedback generation

Meaning group have emerged and formed with small number of student with training data, not computational intensive scalability with large number of data. For the first trial presented in Sect. 5, the ESDNN was trained with the responses for 10 MCQs on English Grammar obtained from previous cohort of students. After training, appropriate feedback text was written by academics for each of the group of students' responses that address the conceptual errors implicit in combinations of incorrect answers. During the trial, a current cohort of students was asked to provide responses on the same questions, they were given the feedback on the combination of incorrect answers and their responses recorded in the database. The feedback texts are composed around the pattern groupings and are aimed at misconceptions that may have caused the incorrect answers common within the pattern group.

An example of a typical response to the questions below is (1<D>2< B>3<A>4<B>5<B/D>6<A>7<A>8<C>9<A/C>10<B/D

1. ______ no cause for alarm, the old man went back to his bedroom.
A. There was *B. Since* *C. Being* ***D. There being***

2. Even as a girl,______ to be her life, and theater audiences were to be her best teachers.
A. performing by Melissa were.
B. it was known that Melissa's performances were
C. knowing that Melissa's performances were
D. Melissa knew that performing was

3. Agriculture is the country's chief source of wealth, wheat ______ by far the biggest cereal crop.
A. is *B. been* *C. be* *D. being*

4. This company has now introduced a policy_____ pay rises are related performance at work.
A. which ***B. where*** *C. whether* *D. What*

5. She managed to save ______she could out of her wages to help her brother.
A. how little money ***B. so little money***
C. such little money *D. what little money*

6. He left orders that nothing______ touched until the police arrived here.
A. should be *B. ought to be* *C. must be* *D. would be*

7. As it turned out to be a small house party, we _________ so formally.
A. need not have dressed up *B. must not have dressed up*
C. did not need to dress up *D. must not dress up*

8. I ______ the party much more if there hadn't been quite such a crowd of people there.
A. would enjoy *B. will have enjoyed*
C. would have enjoyed *D. will be enjoying*

9. There ought to be less anxiety over the perceived risk of mountain climbing than _____ in the public mind today.
A. exists *B. exist* ***C. existing*** *D. to exist*

10. Fat cannot change into muscle ______ muscle changes into fat.
A. any more than ***B. no more than*** *C. no less than* *D. much more than*

This is classified into Group 6, which generates the following feedback:

Group 6 Feedback

Four points should be stressed. First, the logical subject of the adverbial phrase should agree with that of the main clause. Second, two verbs in a sentence need a conjunction. E. G. I am a teacher but you are a student. Third, the usage of various noun clauses should be familiar with. E. G. The news came that he died. ("that" does not serve as any part of the clause.) Fourth, some fixed structures in the comparative form should be memorized. E. G. not so……as……

In total data of six trials were collected (3 English trials, 2 Math trials, 1 Plagiarism awareness trial). In this paper, it presents the details of data collection of 2nd English trial. The data for training is collected from three previous year's MCQs tests. For these three tests, 94 students' answers were used to training. The trials data were collected during academic year 2010–2011. The data of two separate MCQ paper tests and final examination results were gathered. 83 students entered the survey and 16 students were randomly selected for interview. The states of knowledge of students were achieved by using ESDNN.

4.2 English experiments

To investigate and evaluate how the M-OFS guide and support students to learn, three English experiments were under taken by level 2 and level 3 students at JinQiao University (JQU) and Kunming Technology University (KTU) in China during the academic year 2010–2011. The 1st experiment is introduced below.

In the first experiment, data was collected from 148 students taking English language courses whom were randomly separated into two groups. The experimental group of 83 students used M-OFS, and the control group of 65 students received the same training but without using M-OFS. The system trial includes 10 MCQs with 4 potential answers, related to English grammar. The duration of this trial is flexible. When students were using M-OFS, they were encouraged to answer the MCQs (submit their answers) as many times as they wish until they got all the correct answers or gave up (students were not given answers or how many answers were correct in their feedback, except that they answered all correct answers). Two MCQ paper tests with different questions from system trials were applied to 116 students, and 83 students participated in both paper test and system trial. 83 students completed survey after second paper test. This system trial, paper test and survey were completed in practice lessons in a computer room at JQU.

5. EMPIRICAL STUDY

This section discusses the results from the first experiment in order to evaluate the effectiveness of using M-OFS to support students' deep learning.

The survey and interview were conducted after the system trial. 83 (100 %) students conducted the survey. 16 (19 %) students were randomly chosen for interview. For the survey, 71.1 % students are satisfied with using system. 84.4 % students think the feedback is what they need. Using M-OFS to learn were positively evaluated by students, illustrate that the hypotheses H1 is supported. 90.4 % students would like to use the system again. 92.8 % students would like to recommend the system to a friend or classmate in the future. 81.9 % students have never used similar system before. For interviews, most students (94 %) feel this system is useful and helps them to improve their knowledge, it indicates the hypotheses H1 is supported as well; moreover, 69 % students want the exact answers in the feedback in the end. Students also want a picture of their learning process which can point out their weakness and a suggestion of how to improve their English. Some students feel that if they tried many times but cannot find the correct answer, they will lose patience in the end.

5.1 Experiment and result

148 students are involved in the first experiment. 116 students completed the separate MCQs paper test before and after using the system. 83 students participated in system trial, and separate MCQs paper tests. For system trials, a total of 1,118 answers/attempts were submitted and a total 2,143 min were spent by 83 participants. All of the students submitted their answers at least once. The maximum number of attempts was 106 times and the minimum was 1. The average attempts for each student is 13.5 times. The average time spent by each student is 25.8 min and the average time of each attempt is 1.92 min. Two students (2.4 %) spent more than 60 min. 35 (42.2 %) students spent more than the average time. No students achieved the all correct answers at the beginning. 55 (66.3 %) students increased their scores by an average of 12.8 %, whilst one student increased his score by 70 %. In this trial, with 10 questions and 4 possible answers, there are more than one million possible combinations of answers, thus the students are unlikely to make improvement by guessing answers; hence, the results show the feedback had a positive impact which partially supports hypotheses H2. For separate MCQs paper tests, the average score before system trial is 51.6 %, and the average score after system trial is 59.1 %. One student increased his score by 40 %. 74 % students increased their scores. In this test, the students were not given any answers or feedback between first (before system trial) and second (after system trial) test; furthermore, the first trial were applied 3 h before

the system trial and the second test were conducted 30 min after system trial; hence, the students are only learnt by using M-OFS but not any other ways; thus the results above are confident, therefore partially supporting hypotheses H3. In addition, this result also can partially support hypotheses H2. For final examination, both the experimental group and the control group enter the same 4 days final examination. The experimental group got 79.52 % and control group got 71.3 % in English grammar module. This result confirms the hypotheses H4; furthermore, it also supports hypotheses H2.

5.2 Some group behavioural characteristics

Previous work has made an initial investigation of the behavioural characteristics of students during their learning interaction with a diagnostic feedback system Alemán et al. (2011). In order to explore the characteristics of students, and relate these to student responses and performance in the tests, five behavioural variables were analysed: the number of attempts (submissions), the average score changed between attempts, the average score at the end of trial, the amount of time spent to make each attempt, and the learning duration. Figure 2 illustrates a learning behaviour of this group of students by analysing the relationship between average scores increased and learning duration. Each blue point represents average scores increased of all students used the same learning time, and its coordinate of xx-axis represents student's learning duration, and its coordinate of yy-axis represents average scores increased. It can be achieved from this figure that average scores increased when students spent more time on studying from the system.

Figure 3 illustrates a learning behaviour of this group of students by analysing the relationship between average scores increased and number of attempts. Each blue point represents average scores increased of all students did the same number of attempts, and its coordinate of xx-axis represents number of students' attempts, and its coordinate of yy-axis represents average scores increased. It can be achieved from this figure that average scores increased when students did more attempts before peak, and average scores no longer increased when students did 26 attempts, and average scores decreased by doing more attempts after peak.

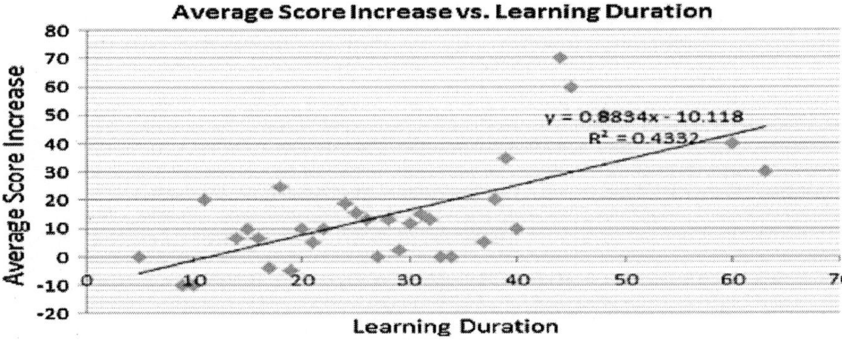

Figure 2. Average score increased versus learning duration

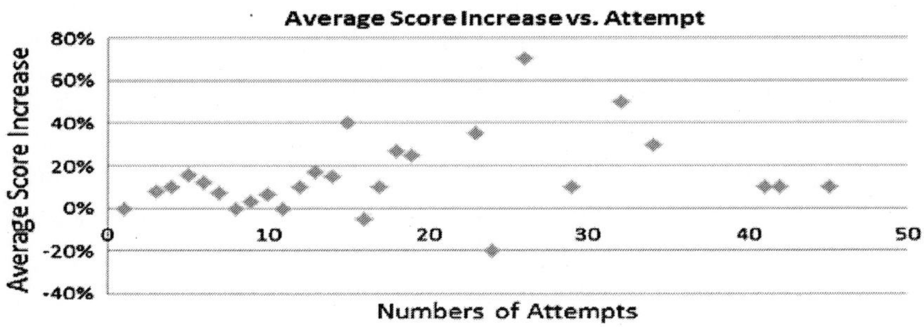

Figure 3. Average score increase versus attempt

Long Learning Duration (time spent on learning ≥≥ 44 min) and Medium Attempts (35≥numberofattempt≥1835≥numberofattempt≥18) are consistently associated with good score increases, especially the combination of Long Learning Duration and Medium Attempts, and hence represent successful learning strategies amongst the students (Tables 3, 4, 5, 6).

Table 3. *Short learning duration*: time spent on learning << 44 min; *Long learning duration*: time spent on learning ≥≥44 min

Behavioral group	Average score change (%)	Average score in the end (%)	Number of students
Short learning duration	9.2	50.6	77
Long learning duration	46.7	95	6

Table 4. *Many attempts*: number of attempt >35>35, *Medium attempts*: 35≥numberofattempt≥1835≥numberofattempt≥18, *Few attempts*: number of attempt << 18

Behavioral group	Average score change (%)	Average score in the end (%)	Number of students
Many attempts	8.8	51.2	65
Medium attempts	31.5	72.3	13
Few attempts	15	57.5	4

Table 5. *Slow attempts*: average time spent on each attempt ≥≥ 4 min; *Rapid attempts*: average time spent on each attempt <4<4 min

Behavioral group	Average score change (%)	Average score in the end (%)	Number of students
Slow attempt	12.96	53.52	54
Rapid attempt	11.72	54.48	29

Table 6. *Few rapid attempts*: average time spent on each attempt << 4 min and number of attempt << 18; *Many rapid attempts*: number of attempt >>35 and average time spent on each attempt << 4 min; *Slow Few attempts*: average time spent on each attempt ≥≥ 4 and number of attempt << 18; *Rapid medium attempts*: 35≥numberofattempt≥ 1835≥ numberofattempt≥18 and average time spent on each attempt << 4 min

Behavioral group	Average score change (%)	Average score in the end (%)	Number of students
Few rapid attempts	6	48.4	45
Many rapid attempts	15	57.5	4
Slow few attempts	16	54.5	20
Rapid medium attempts	31.5	72.3	13

Figure 4 illustrates the behaviour of students in terms of what might be called knowledge states. These states correspond to the student responses triggered by patterns of student answer. In other words, a state of knowledge captures some commonality in a set of questions responses. For example, if there are several students who give the same answer (correct or incorrect) to two or more of the questions, snap-drift will form a group associated with one particular output neuron to include all such cases. That is an over simplification, because some of those cases may be pulled in to other 'stronger' groups, but that would also be characterized by a common feature amongst the group of responses. Figure 3 shows the knowledge state transitions. Each time a student gives a new set of answers, having received some feedback associated with their previous state, which in turn is based on their last answers, they are reclassified into a new (or the same) state, and thereby receive new (or the same) feedback. The tendency is to undergo a state transition immediately or after a second attempt or several attempts. A justification for calling the states 'states of knowledge' is also to be found in their self-organization into the layers. There are 5 layers: Start, Layer 1, Layer 2, Layer 3, and Layer 4. A student on state 5, for example goes via one of the states in the further layer such as state 4 or 13, before reaching the 'state of perfect knowledge' (state 20) which represents correct answers to all questions. On average, and unsurprisingly, the state-layer projecting onto state 20 (states 12) is associated with more correct answers than the states in the previous layer. This is true of state 13 which projects onto state 20, and it is also true of state 4 although this state does not project onto state 20. The states in the middle layer (layer 2) do not connect to start layer or layer 4. The average score of each layer is increased from start to state 20 (Average scores are 69.38 % at beginning level (layer 1), 73.33 % at middle layer (layer 2), 81 % at advanced level (layer 3), and 100 % at state 20). Students often circulate within layers before proceeding to the next layer. They may also return to a previous layer, but that is less common.

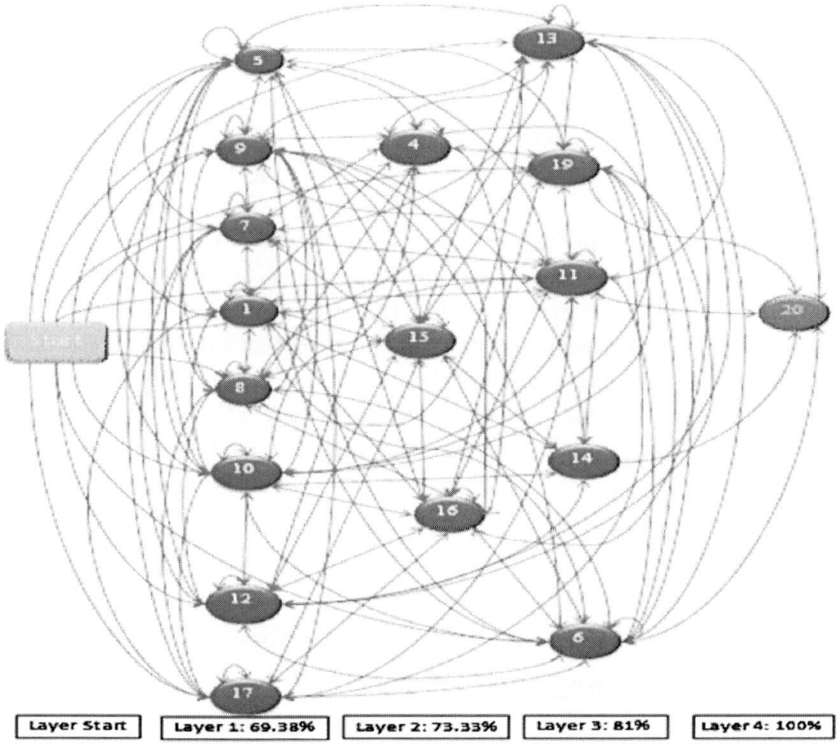

Figure 4. Knowledge state transitions

6. SUMMARY

Six trials across three totally different subject areas have been carried out in three universities in two countries, the UK and China: 3 English trials, 2 Mathematics trials, and 1 Java Programming trial. The English trials are very successful with 500 students participated. A large volume of data has been captured during the trials.

The students participating in this second trial are from KTU which is one of the top universities in China. The entrance requirements of KTU are much higher than those of JQU, which means that students studying in KTU generally have a better academic background than students in JQU. The results of the second English system trials show that the average score is increased by 21.2 % in the end. The results of the separate MCQ paper test indicated the average mark of the group is increased by 7.6 %. In addition, the final examination, the average mark of the experimental group is 7.5 % higher than the control group. Student surveys show that 87.8 % students are satisfied with our e-learning system and 89.2 % students feel that the intelligent diagnostic feedback is beneficial to enhance their learning experience. Furthermore, according to all of the results, the M-OFS system demonstrates that it is working well for both less qualified and better qualified students at two universities JQU and KTU. The results of group behaviours in Table 7 show that students in the group "Long learning duration medium attempts" achieved much better learning effects than others. In comparison with the 1st trial, the students in the group "Long learning duration medium attempts" achieved better learning effects than others. This has indicated that the students in the 2nd trial who have better academic background prefer more reflective thinking and independent study from the feedbacks, whilst, the students in the 1st trial who have weaker academic background prefer more frequent and direct feedbacks as part of their learning process.

7. CONCLUSION AND FUTURE WORK

This research develops a novel method for using snap-drift in a diagnostic tool to provide feedback. The neural network diagnostic feedback approach in MCQ has been systemically applied to large cohorts of students and evaluated across a range of different subject areas. The neural network discovers groups of similar answers that represent different knowledge states of the students. The feedback targets the level of knowledge of individuals, and their misconceptions, guiding them toward a greater understanding of particular concepts. The results of the experiment show that an improvement in the learning process can be obtained by using M-OFS. In future work, it is intended to compare the effects of the feedback in new trials to the effects of other types of feedback already reported in the literature. Another promising avenue for further investigation is the extension of the tool to support knowledge state transition diagram construction and statistical data collection, which could help instructors to analyse the difficulty of the MCQs and to track students though the developmental stages of their learning.

Table 7. *Short learning duration many attempts*: time spent on learning << 44 min and number of attempt >>35; *Short learning duration medium attempts*: time spent on learning << 44 min and 35≥35≥ number of attempt ≥18≥18;*Short learning duration few attempts*: time spent on learning << 44 min and number of attempt << 18; *Long learning duration few attempts*: time spent on learning ≥≥ 44 min and number of attempt << 18; *Long learning duration medium attempts*: time spent on learning ≥≥ 44 min and 35≥35≥ number of attempt ≥18≥18

Behavioral group	Average score change (%)	Average score in the end (%)	Number of students
Short learning duration few attempts	7.9	48.4	62
Short learning duration many attempts	16	57.5	4
Short learning duration medium attempts	23	64	10
Long learning duration few attempts	33.3	90	3
Long learning duration medium attempts	60	100	3

REFERENCES

1. Alemán JLF, Palmer-Brown D, Jayne C (2011) Effects of response-driven feedback in computer science learning. IEEE Trans Educ 54:501–508

2. Alessi M, Trollip S (2001) Multimedia for learning: methods and development, 3rd edn. Allyn and Bacon, London

3. Bang P (2003) Engaging the learner—how to author for best feedback. In: Felix U (ed) Language learning online: towards best practice. Swets & Zeitlinger, The Netherlands

4. Blessing S, Gilbert S, Ourada S, Ritter S (2007) Lowering the bar for creating model-tracing intelligent tutoring systems. In: Luckin R et al. (eds) Artificial intelligence in education. IOS Press, Amsterdam

5. Clark RC, Mayer RE (2009) Instructional strategies for directive learning environments. In: Silber KH, Foshay WR (eds) Handbook of improving performance in the workplace: instructional design and training delivery. Pfeiffer, San Francisco

6. Epstein ML, Lazarus AD, Calvano TB, Mathews KA, Hendel RA, Epstein BB, Brosvic GM (2002) Immediate feedback assessment technique promotes learning and corrects inaccurate first response. Psychol Record 52(2):187–201

7. Garber PR (2004) Giving and receiving performance feedback. HRD Press, Canada

8. Gheorghiu R, Vanlehn K (2008) XTutor: an intelligent tutor system for science and math based on excel. In: Woolf BP et al. (eds) Intelligent tutoring systems: 9th international conference, ITS 2008. Springer, Germany

9. Hatziapostolou T, Paraskakis I (2010) Enhancing the impact of formative feedback on student learning through an online feedback system. Electron J E-Learn 8(2):111–122. Available online at www.ejel.org

10. Hefce national student survey (2007–2010) HEFCE, London, UK. Online at: http://www.hefce.ac.uk/whatwedo/lt/publicinfo/nationalstudentsurvey/

11. Heinze A, Procter C, Scott B (2007) Use of conversation theory to underpin blended learning. Int J Teach Case Stud 1(1 &2):108–120

12. Higgins E, Tatham L (2003) Exploring the potential of multiple-choice questions in assessment. Learn. Teach. Action 2(1). Assessment, Winter 2003

13. Kuechler WL, Simkin MG (2003) How well do multiple choice tests evaluate student understanding in computer programming classes? J Inf Syst Educ 14(4):389–399

14. Little B (2001) Achieving high performance through e-learning. Ind Commer Train 33(6):203–206

15. Ma Z (2006) Web-based intelligent e-learning systems. Information Science Publishing, USA

16. Nelson MM, Schunn CD (2009) The nature of feedback: how different types of peer feedback affect writing performance. Instr Sci 37(4):375–401

17. Özdener N, Satar HM (2009) Effectiveness of various oral feedback techniques in CALL vocabulary learning materials. Egitim Arastirmalari—Eurasian J Educ Res 34:75–96

18. Palmer-Brown D, Jayne C (2011) Snap-drift neural network for self-organisation and sequence learning. Neural Netw 24(8):897–905

19. Paxton M (2000) A linguistic perspective on multiple-choice questioning assessment and evaluation. Assess Eval High Educ 25(2):109–110

20. Payne A, Brinkman, Wilson F (2007) Towards effective feedback in e-learning packages: The design of a package to support literature searching, referencing and avoiding plagiarism. In: Proceedings of HCI2007 workshop: design and use and experience of e-learning systems, pp 71–75

21. Race P, Brown S (2005) 500 tips for tutors, 2nd edn. Routledge Falmer, London

22. Race P (2006) The lecturer's toolkit—a practical guide to assessment. Learning and teaching, 3rd edn. Routledge, London

23. Rane A, Sasikumar M (2007) A constructive learning framework for language tutoring. In: Iskander M (ed) Innovations in e-learning, instruction technology, assessment, and engineering education. Springer, The Netherlands

24. Reiser BJ, Kimberg DY et al (1992) Knowledge representation and explanation. In: Larkin JH, Chabay RW (eds) Computer assisted instruction and intelligent tutoring systems. Lawrence Erlbaum Associates, New Jersey

25. Springgay S, Clarke A (2007) Mid-course feedback on faculty teaching: a pilot project. In: Darling LF, Erickson GL, et al (eds) Collective improvisation in a teacher education community, Chapter 13. Springer, The Netherlands, pp.171–185

26. Vasilyeva E, Pechenizkiy M, Bra PD (2008) Adaptation of elaborated feedback in e-learning. In: Nejdl W, Kay J, Pu P (eds) Adaptive hypermedia and adaptive web-based systems. Springer, German

Index

Index